Biogeographic Patterns of South American Anurans

Tiago S. Vasconcelos • Fernando R. da Silva
Tiago G. dos Santos • Vitor H. M. Prado
Diogo B. Provete

Biogeographic Patterns of South American Anurans

 Springer

Tiago S. Vasconcelos
Department of Biological Sciences
São Paulo State University (UNESP)
Bauru, São Paulo, Brazil

Tiago G. dos Santos
Federal University of Pampa (UNIPAMPA)
São Gabriel, Rio Grande do Sul, Brazil

Diogo B. Provete
Federal University of Mato Grosso do Sul
(UFMS)
Campo Grande, Mato Grosso do Sul, Brazil

Fernando R. da Silva
Federal University of São Carlos (UFScar)
Sorocaba, São Paulo, Brazil

Vitor H. M. Prado
Goiás State University (UEG)
Anápolis, Goiás, Brazil

ISBN 978-3-030-26298-3   ISBN 978-3-030-26296-9   (eBook)
https://doi.org/10.1007/978-3-030-26296-9

This Springer imprint is published by the registered company Springer Nature Switzerland AG
The registered company address is: Gewerbestrasse 11, 6330 Cham, Switzerland

Foreword

Biogeography is, in its origin and definition, a multidisciplinary science that encompasses different branches of knowledge, such as community ecology, evolutionary biology, genetics, geology, paleontology, and climatology. Thus, only gathering a team like this, made up of researchers with solid theoretical background and different expertise, coupled with the natural disposition and energy of the young people, this book could be elaborated. They were courageous enough to analyze several aspects of the distribution of anurans in South America in a broad and at the same time in-depth approach.

This is the first update of species richness and the first analysis of anuran distribution patterns across South America since 1999, when the first synthesis developed by W. E. Duellman ("Distribution Patterns of Amphibians in South America") was published. However, rather than simply expanding the documentation and mapping of geographic distribution of more than 2000 species, the authors innovate by documenting and mapping different components of anuran diversity, such as the mapping of phylogenetic diversity and several other metrics of functional diversity throughout the continent. Moreover, they explore evolutionary and ecological processes and indicate, based on this broad set of analyses, priority areas for conservation and what is the minimum conservation area necessary to maintain the diversity of anurans in the continent.

Despite the extensive and in-depth analysis, reading this book is extremely enjoyable due to its pleasant language, accessible not only to researchers but also to students and researchers unfamiliar with this subject, who may greatly benefit from the knowledge synthesized here. Inserting studies in areas as diverse as genetics, molecular biology, and phylogeography in a biogeographic and evolutionary context makes it possible to take current knowledge a step further.

The lightness placed in this book reveals a work done with pleasure, fruit of the joy in knowing how the nature works and how knowledge is produced. But it also originates from the pleasure of working with friends, as is the case of the authors who have spent, at different periods and for different time spans, part of their training as scientists at the Laboratory of Theoretical Ecology of the UNESP (São Paulo State University, Brazil). Nowadays, when personal opinions are having the same value of scientific knowledge, which has been brutally disregarded by part of the population, reading this book brings hope and encouragement! Earth is not flat, vaccines are fundamental, and universities, scientists, and teachers at all school levels are fundamental and indispensable if we want to live in a more fair, prosperous, and healthy society!

São José do Rio Preto, Brazil   Denise de C. Rossa-Feres
April 2019

Preface

We all met each other sharing the personal interest of studying anuran communities from Southeastern Brazil, specifically, anurans from the northwestern region of the São Paulo State, Brazil. The institution was the campus of São José do Rio Preto of the São Paulo State University (UNESP, SP, Brazil), where our supervisor was Prof. Dr. Denise de Cerqueira Rossa-Feres. Some of us (TSV and VHMP) started earlier during our scientific initiation projects required for our graduation in the course of Biological Sciences, whereas the three of us (DBP, FRdS, and TGdS) arrived later in the Laboratory of Theoretical Ecology of the UNESP for the master's degree in Animal Biology of the same institution. Though strongly focused on field ecology of anuran communities, the experience in the laboratory led by Dr. Rossa-Feres allowed us a working environment and stimulation for everyone to explore community ecology in its broadest senses: local to broad-scales, natural history to statistically- designed approaches. Dr. Rossa-Feres guided us during the early steps of our scientific careers, and, besides all benefits from the professional experiences, we also built strong friendship among one other. All that said, we want to dedicate this book to Prof. Dr. Denise Rossa-Feres, our common supervisor and example of dedication in scientific life.

Despite our strong background on community ecology focused on anuran distributions at local scales (i.e., breeding ponds), biogeography is a discipline that has called our attention since our early days in science (early 2000s). While studying the classical references of William E. Duellman, one of them served us to better understand the patterns of anuran distribution in Southeastern Brazil, the book chapter "Distribution Patterns of Amphibians in South America" (published in 1999 by the John Hopkins University Press). By the mid- 2016, we realized that a plenty of biogeographical studies considering anurans in South America were performed since 1999, but a systematic quantification of how many species are documented in South America was missing. Moreover, we realized that we could perform different analyses to quantify and map different biological diversity metrics that were never or seldom explored in the literature so far. Then, we planned to update the anuran species list in South America and, based on our different expertise, split the major biogeographic themes so that each one of us could contribute chapters of this book.

Then, if the readers are interested in a particular subject of a specific chapter, we recommend them to contact the specific author that was the main responsible for delineating and performing the main analyses of the respective chapter: TS Vasconcelos: Chaps. 1, 6, and 7; FR da Silva: Chap. 4; TG dos Santos: Chap. 2; VHM Prado: Chap. 3; and DB Provete: Chap. 5.

Writing up this book involved a great amount of work that would not be possible without the assistance of colleagues and/or students that helped us in gathering, tabulating, and checking excel files, among others, of the great amount of information processed. For this, we are grateful to Brena Gonçalves Silva (UNIPAMPA, São Gabriel, Brazil), Guilherme Castro Franco (UNIPAMPA, São Gabriel, Brazil), and Phillip T. Soares, MSc (UFMS, Campo Grande, Brazil). We are also grateful to the following colleagues who kindly served as external reviewers and for contributing to the quality of this book: Prof. Dr. Igor Luis Kaefer (Universidade Federal do Amazonas, Manaus, Brazil, reviewed Chap. 2), Prof. Dr. Fabrício Barreto Teresa (Universidade Estadual de Goiás, Campus Anápolis de Ciências Exatas e Tecnológicas "Henrique Santillo," Anápolis, Brazil, reviewed Chap. 3), Prof. Dr. Victor Satoru Saito (Universidade Federal de São Carlos, Departamento de Ciências Ambientais, São Carlos, Brazil, reviewed Chap. 4), Prof. Dr. Bruno Vilela de Moraes e Silva (Universidade Federal da Bahia, Salvador, Brazil, reviewed Chap. 5), and Prof. Dr. Peter Löwenberg-Neto (Universidade Federal da Integração Latino-Americana, Foz do Iguaçu, Brazil, reviewed Chap. 6).

Finally, we are also grateful to the following research agencies that have been continuously supporting our research activities in Brazil: Fundação de Amparo à Pesquisa do Estado de São Paulo (FAPESP grants n°: 2011/18510-0; 2013/50714-0; 2016/13949-7), Conselho Nacional de Desenvolvimento Científico e Tecnológico (CNPq, grants n°: 2037/2014-9; 431012/2016-4; 308687/2016-17; 114613/2018-4), and the University Research and Scientific Production Support Program of the Universidade Estadual de Goiás.

Bauru, São Paulo, Brazil   Tiago S. Vasconcelos
Sorocaba, São Paulo, Brazil   Fernando R. da Silva
São Gabriel, Rio Grande do Sul, Brazil   Tiago G. dos Santos
Anápolis, Goiás, Brazil   Vitor H. M. Prado
Campo Grande, Mato Grosso do Sul, Brazil   Diogo B. Provete

Contents

Chapter 1
An Introduction to the Biogeography of South American Anurans

Abstract South America has undergone complex environmental and geological events that ultimately made it the most climatically and biodiverse continent on the planet, including anuran amphibians. Though biogeographical studies with anurans in South America have been continuously performed during the last decades, most of them focus on specific clades and/or regions. Moreover, no systematic compilation has been performed since the first synthesis of patterns of amphibian distribution, conducted by William E. Duellman by the end of the twentieth century. Here, we perform a systematic species survey of anurans in South America that allow us to revisit previously documented biogeographic patterns of species distribution (e.g., geographical patterns of species richness, species range size) and uncover novel biogeographic patterns, such as mapping anuran phylogenetic and functional diversity metrics across the continent. In summary, this book is made up of seven chapters spanning a wide range of topics that integrate herpetology, biogeography, ecology, and conservation biology. This chapter provides an overview of South American anurans and details the methodology used to generate the species list used in all subsequent chapters, as well as how different environmental variables were gathered and processed to be assessed as potential predictors of the biodiversity metrics explored through this book.

Keywords Anura · Biogeography · Lissamphibia · Macroecology · Neotropics · South America

1.1 Introduction

The New World comprises the continents and their associated islands of the planet's western hemisphere. The western continents were formed after the complex breakup of the Pangea supercontinent, which happened from the Late Triassic (~220 Mya) to the Late Cretaceous (~80 Mya) (Lomolino et al. 2017 and references therein). Specifically, South America was part of the southern landmasses of Pangea (the

© Springer Nature Switzerland AG 2019
T. S. Vasconcelos et al., *Biogeographic Patterns of South American Anurans*,
https://doi.org/10.1007/978-3-030-26296-9_1

Gondwana) and started splitting apart from the African landmass approximately 100 Mya (Lomolino et al. 2017). Currently ranging from the equatorial region to southern temperate latitudes, South America has undergone a series of complex environmental and geological events that ultimately made it the most climatically and biologically diverse continent on the planet (Rangel et al. 2018). Among these events, it is well recognized that (a) the intensity of Pleistocene glaciation (e.g., the Last Glacial Maximum that occurred about 26,500–19,000 years ago) was unevenly distributed across the continent, creating regions that were climatically more or less stable. Those stable regions acted as refuges and had high rates of intraspecific diversification, as well as high species richness (e.g., Carnaval and Moritz 2008; Carnaval et al. 2009; Lomolino et al. 2017); (b) the uplift of mountain chains favored biological diversification through allopatric speciation, such as the Andes mountain complex uplift in the Neogene (~23–2,5 Mya) (e.g., Ruggiero and Hawkins 2008; Antonelli et al. 2009; Rangel et al. 2018); and (c) after splitting apart from the Gondwana, South America remained isolated from any other landmasses, so in situ lineages diversification took place at this "giant" island for over 50 million years. Then, the uprising of the isthmus of Panama, between 23 and 7 Mya (Bacon et al. 2015), bridged South America to Central and North America, thus favoring the exchange of lineages between regions (Antonelli et al. 2018).

Among the different biological groups, South America is home to the most diverse anuran fauna of the world (Wake and Koo 2018). Anurans are the most diverse order of amphibians, comprising 56 families and approximately 7053 species described worldwide (Frost 2019). Commonly known as frogs, treefrogs, and toads, these animals are scaleless (i.e., they have a highly permeable, bare skin that allows water and gas exchange), tailless, and have highly adapted legs for jumping locomotion (Duellman and Trueb 1994; Wells 2007; Haddad et al. 2013). They were one of the first vertebrate lineages that reached terrestrial environments approximately 200 Mya (Carroll 2009; Stocker et al. 2019 and references therein). Today, anurans are highly diverse in the Tropics of South America, whereas low diversity is mostly found in the temperate region or in tropical high-altitude sites west of the Andes (Wake and Koo 2018).

As the science that documents and understands spatial patterns of biological diversity (Lomolino et al. 2017), biogeography requires a great amount of information from different fields (e.g., population and community ecology, evolutionary biology, genetics) coupled with other natural sciences (e.g., geology, paleontology, climatology) in order to understand the distribution of biodiversity on Earth. Thus, the fundamental unity for biogeographical studies is the geographic distribution of species. For few taxonomic groups, and/or over broad regions (e.g., Europe and North America), distribution information for most species is available, which facilitates biogeographical studies (e.g., Currie and Paquin 1987; Whittaker et al. 2007; Hawkins 2010). In South America, the development of anuran biogeography was only possible with the accumulation and organization of the geographic distribution of species. Then, the first synthesis of patterns of amphibian distribution was performed by the end of the twentieth century (Duellman 1999). This study compiled occurrence data for 1644 anurans in South America and discussed their distribution patterns among and within biogeographic subregions (Duellman 1999). Afterward, anuran biogeographical studies in South America were performed in the context of a variety of studies that combined a range of subdisciplines, yet mainly focused on specific clades and/or regions, such as the historical biogeography of some specific clades (e.g., Wiens et al. 2011; Fouquet et al. 2013), phylogeography (e.g., Carnaval et al. 2009; Gehara et al. 2014), ecological biogeography (e.g., Diniz-Filho et al. 2006, 2008; Vasconcelos et al. 2010; da Silva et al. 2012), and regionalization schemes (e.g., Valdujo et al. 2013; Vasconcelos et al. 2014; Godinho and da Silva 2018) for specific regions in South America. Studies involving the whole continent generally documented species richness gradients and range size patterns of amphibians, irrespective of the specific aims (e.g., Buckley and Jetz 2007; Vasconcelos et al. 2012; Villalobos et al. 2013; Wake and Koo 2018). Other approaches proposed regionalization schemes (Vasconcelos et al. 2011) or mapped phylogenetic diversity metrics (Fritz and Rahbek 2012), but the common share of all these studies considering the continental and/or global background is the use of the same dataset: the extent-of-occurrence polygons elaborated and provided by the International Union for Conservation of Nature (IUCN red list of threatened species: www.iucnredlist. org). Though the superimposition of such species range maps onto a South America grid generally produces more than 2400 anuran species in the continent (e.g., Vasconcelos et al. 2011; Villalobos et al. 2013), no systematic compilation has been performed since Duellman (1999).

Global databases of amphibian information have become an important tool for herpetologists to track the dynamic addition of new species formally described and follow taxonomic changes resulted from systematic reviews (e.g., AmphibiaWeb 2019; Frost 2019; IUCN 2019). In this book, we take advantage of these databases to perform a systematic species survey of anurans in South America that allow us to revisit previously documented biogeographic patterns of species distribution (e.g., geographical patterns of species richness, species range size, and regionalization schemes) and uncover novel biogeographic patterns, such as mapping anuran phylogenetic and functional diversity metrics across the continent.

Specifically, this book is made up of seven chapters spanning a wide range of topics that integrate herpetology, biogeography, ecology, and conservation biology. Chapter 1 provides an overview of South American anurans, connecting the biogeographical history of South America, followed by the methodology used to generate the species list used in all subsequent chapters. Chapter 2 provides the updated anuran species list compiled for South America and discusses trends in the spatiotemporal dynamic of anuran descriptions. Chapters 3, 4, and 5 explore macroecological patterns of different biodiversity metrics. In Chap. 3, we revisit the well-documented species richness and range size patterns, yet we perform cutting-edge statistical approaches to identify the environmental correlates of such ecological metrics. In Chap. 4, we map two phylogenetic diversity indices and perform null-model analysis to look for regions with higher phylogenetic diversity index than expected by their species richness. In Chap. 5, we map multiple dimensions of anuran functional diversity based on an extensive trait database recently available (Oliveira et al. 2017), but also complemented with data from the literature. In Chap. 6, we propose a new regionalization scheme for the South American anuran fauna, with an enhanced and updated dataset compared with previous studies. Additionally, we perform regression analyses to identify the main environmental correlates of the clustering patterns and perform variance partitioning to quantify independent and shared components of these predictors. Finally, Chap. 7 presents a spatial conservation prioritization exercise to identify a minimum set of coverage area that is biologically important for conserving anurans in South America using diversity and human-related metrics to optimize our final conservation proposal.

1.2   General Methodology and Data Processing

1.2.1   Anuran Survey in South America and Range Maps

To produce a species list of anurans occurring in South America, we started by downloading the polygons of extent of occurrence (range maps) of 2297 species from IUCN (2016). Next, we updated the species list by consulting the online catalogs of the Amphibian Species of the World (Frost 2019) and AmphibiaWeb (2019) up to June 2017, and then we added those anuran species that were not included in the IUCN database. This added 421 species to our list. Following the Amphibian Species of the World (Frost 2019), also until June 2017, we checked each 2718 species for nomenclature changes (e.g., synonymies, species revalidations), so this filtering process finally accounted for the anuran species list that is provided in Chap. 2.

Subsequently, we performed a literature survey to gather occurrence records for those added species. When a species had three or more occurrence records, their geographic ranges were generated in ArcGIS 10.1 using the function "minimum bounding geometry." This function generates a polygon considering the shortest distance between any two vertices of the convex hull, a procedure commonly used to generate the extent of occurrence of species (e.g., García-Roselló et al. 2014). For those species with one or two occurrences only, their ranges were considered as the area within ~50 km of diameter of each occurrence record. Finally, using the package letsR in the R environment (Vilela and Villalobos 2015), the species ranges of all species were overlaid onto a grid cell system in South America at 1° resolution (total of 1649 cells), to obtain a presence/absence matrix of species occurrence that was used in all subsequent chapters.

Among a variety of mapping methods of species distributions, three of them are recurrently used by biogeographers (Graham and Hijmans 2006; Hawkins et al. 2008; Vasconcelos et al. 2012; García-Roselló et al. 2014): (a) point-to-grid maps, which usually consider point observations of species occurrences into a pre-defined grid system; (b) extent-of-occurrence range maps, which presume that the distribution area of a species are composed of connected populations, so the distributional area is a typical polygon connecting the occurrence records, and; (c) ecological niche modeling or species distribution model maps, which usually use environmental predictors for model building (generally climatic variables), thus assuming that these environmental variables are the main determinants of species distributions at broad spatial scales. In summary, point-to-grid maps usually underestimate species occurrences ("errors of omission") due to insufficient and spatially biased samplings (e.g., Graham and Hijmans 2006 and references therein). Conversely, overestimations of species ranges are usually recorded in extent of occurrence and, even higher, in distribution modeling maps, which is mostly related to the consideration of areas in between the occurrence records (Graham and Hijmans 2006). Considering this continuum of "commission – omission errors," the extent-of-occurrence range method seems to be in an intermediate position compared to the point-to-grid and modeling approaches. Then, although having some level of error (Ficetola et al. 2014), we chose the IUCN (2016) range maps as the main source of anuran distribution in South America. Within a biogeographical perspective (e.g., the 1° resolution considered herein), range maps can be as accurate as point occurrence records at grains greater than 50 km if the intent is to document broad-scale biodiversity patterns (Hawkins et al. 2008).

1.2.2   Environmental Data

We used different environmental variables (i.e., climatic, topographic, human-related, vegetation, and habitat structure variables) to be assessed as potential predictors of the biodiversity metrics explored through this book, specifically in Chapters 3, 6, and 7. All quantitative environmental variables were averaged for each cell of the South America grid system, whereas the qualitative habitat structure variable (see ahead) was considered as the most dominant major biome within each grid cell.

The following climatic variables were obtained from the WorldClim v. 1.4 at a resolution of ~10 km (Hijmans et al. 2005): average annual temperature, temperature seasonality, annual precipitation, and precipitation seasonality. Annual actual evapotranspiration, a measure of water-energy balance, was obtained at the same ~10 km resolution from http://www.fao.org/geonetwork/srv/enn/metadata.show?currTab=simple&id=52366.

The elevation range within each grid cell (i.e., the subtraction of the maximum by the minimum altitude value), a measure of topographic heterogeneity, was calculated based on the altitude data (~1 km resolution) available at https://lta.cr.usgs.gov/GTOPO30.

The normalized difference vegetation index (NDVI), a measure of primary productivity, was obtained from the Global Inventory Modeling and Mapping Studies (http://www.glcf.umd.edu/data/ndvi/) at the resolution of ~10 km. Mean canopy height (CANOP) and standard deviation of canopy height (CANSD), two measures of vegetation complexity, were obtained at 1.0 km resolution from the 3D Global Vegetation Map database (Simard et al. 2011). We also characterized the major habitat structure of each grid cell according to its major biome from the World Wildlife Fund (Olson et al. 2001). Finally, the human footprint index, a combination of negative anthropogenic impacts on the environment (WCS and CIESIN 2005), was downloaded at a resolution of ~1 km.

Acknowledgments The authors have been continuously supported by research grants and/or fellowships from the Fundação de Amparo à Pesquisa do Estado de São Paulo (FAPESP 2011/18510-0; 2013/50714-0; 2016/13949-7), Conselho Nacional de Desenvolvimento Científico e Tecnológico (CNPq 2037/2014-9; 431012/2016-4; 308687/2016-17; 114613/2018-4), and University Research and Scientific Production Support Program of the Goias State University (PROBIP/UEG).

References

AmphibiaWeb (2019) University of California, Berkeley. https://amphibiaweb.org. Accessed 27 Mar 2019

Antonelli A, Nylander JAA, Persson C et al (2009) Tracing the impact of the Andean uplift on Neotropical plant evolution. PNAS 106:9749–9754. https://doi.org/10.1073/pnas.0811421106

Antonelli A, Zizka A, Carvalho FA et al (2018) Amazonia is the primary source of Neotropical biodiversity. PNAS 115:6034–6039. https://doi.org/10.1073/pnas.1713819115

Bacon CD, Silvestro D, Jaramillo C et al (2015) Biological evidence supports an early and complex emergence of the Isthmus of Panama. PNAS 112:6110–6115

Buckley LB, Jetz W (2007) Environmental and historical constraints on global patterns of amphibian richness. P Roy Soc B-Biol Sci 274:1167–1173

Carnaval AC, Moritz C (2008) Historical climate modelling predicts patterns of current biodiversity in the Brazilian Atlantic forest. J Biogeogr 35:1187–1201

Carnaval AC, Hickerson MJ, Haddad CFB et al (2009) Stability predicts genetic diversity in the Brazilian Atlantic Forest hotspot. Science 323:785–789

Carroll R (2009) The rise of amphibians: 365 million years of evolution. The Johns Hopkins University Press, Baltimore

Currie DJ, Paquin V (1987) Large-scale biogeographical patterns of species richness of trees. Nature 329:326–327

da Silva FR, Almeida-Neto M, Prado VHM et al (2012) Humidity levels drive reproductive modes and phylogenetic diversity of amphibians in the Brazilian Atlantic Forest. J Biogeogr 39:1720–1732

Diniz-Filho JAJ, Bini LM, Pinto MP et al (2006) Anuran species richness, complementarity and conservation conflicts in Brazilian Cerrado. Acta Oecol 29:9–15. https://doi.org/10.1016/j.actao.2005.07.004

Diniz-Filho JAF, Bini LM, Vieira CM et al (2008) Spatial patterns of terrestrial vertebrates species richness in the Brazilian Cerrado. Zool Stud 47:146–157

Duellman WE (1999) Distribution patterns of amphibians in South America. In: Duellman WE (ed) Patterns of distribution of amphibians. The Johns Hopkins University Press, Baltimore/London, pp 255–327

Duellman WE, Trueb L (1994) Biology of amphibians. The John Hopkins University Press, Baltimore

Ficetola GF, Rondinini C, Bonardi A et al (2014) An evaluation of the robustness of global amphibian range maps. J Biogeogr 41:211–221. https://doi.org/10.1111/jbi.12206

Fouquet A, Cassini CS, Haddad CFB et al (2013) Species delimitation, patterns of diversification and historical biogeography of the Neotropical frog genus *Adenomera* (Anura, Leptodactylidae). J Biogeogr 41:855–870

Fritz SA, Rahbek C (2012) Global patterns of amphibian phylogenetic diversity. J Biogeogr 39:1373–1382

Frost DR (2019) Amphibian species of the world: an online reference. Version 6.0. American Museum of Natural History, New York. http://research.amnh.org/herpetology/amphibia/index. html. Accessed 27 Mar 2019

García-Roselló E, Guisande C, Manjarréz-Hernández A et al (2014) Can we derive macroecological patterns from primary Global Biodiversity Information Facility data? Glob Ecol Biogeogr 24(335–347):2014

Gehara M, Crawford AJ, Orrico VGD et al (2014) High levels of diversity uncovered in a widespread nominal taxon: continental phylogeography of the Neotropical tree frog *Dendropsophus minutus*. PLoS One 9:e103958

Graham CH, Hijmans RJ (2006) A comparison of methods for mapping species range and species richness. Glob Ecol Biogeogr 15:578–587

Godinho MBC, da Silva FR (2018) The influence of riverine barriers, climate, and topography on the biogeographic regionalization of Amazonian anurans. Sci Rep 8:3427. https://doi. org/10.1038/s41598-018-21879-9

Haddad CFB, Toledo LF, Prado CPA et al (2013) Guide to the amphibians of the Atlantic Forest: diversity and biology. Anolis Book, Sao Paulo

Hawkins BA (2010) Multiregional comparison of the ecological and phylogenetic structure of butterfly species richness gradients. J Biogeogr 37:647–656

Hawkins BA, Rueda M, Rodriguez MA (2008) What do range maps and surveys tell us about diversity patterns? Folia Geobot 43:345–355

Hijmans RJ, Cameron SE, Parra JL et al (2005) Very high resolutions interpolated climate surfaces for global land areas. Int J Climatol 25:1965–1978. https://doi.org/10.1002/joc.1276

IUCN (2016) The IUCN red list of threatened species. Version 2016-1. http://www.iucnredlist.org. Accessed 30 Nov 2016

IUCN (2019) The IUCN red list of threatened species. Version 2019-1. http://www.iucnredlist.org. Accessed 21 Mar 2019

Lomolino MV, Riddle BR, Whittaker RJ (2017) Biogeography: biological diversity across space and time, 5th edn. Sinauer Associates Inc, Sunderland

Oliveira BF, Sao-Pedro VA, Santos-Barrera G et al (2017) AmphiBIO, a global database for amphibian ecological traits. Sci Data 4:170123. https://doi.org/10.1038/sdata.2017.123

Olson DM, Dinerstein E, Wikramanayake ED et al (2001) Terrestrial ecoregions of the world: a new map of life on earth. Bioscience 51:933–938

Rangel TF, Edwards NR, Holden PB et al (2018) Modeling the ecology and evolution of biodiversity: biogeographical cradles, museums, and graves. Science 361:eaar5452. https://doi. org/10.1126/science.aar5452

Ruggiero A, Hawkins BA (2008) Why do mountains support so many species of birds? Ecography 31:306–315. https://doi.org/10.1111/j.2008.0906-7590.05333.x

Simard M, Pinto N, Fisher JB et al (2011) Mapping forest canopy height globally with spaceborne lidar. J Geophys Res-Biogeo 116:G04021

Stocker MR, Nesbitt SJ, Kligman BT et al (2019) The earliest equatorial record of frogs from the Late Triassic of Arizona. Biol Lett 15:20180922. https://doi.org/10.1098/rsbl.2018.0922

Valdujo PH, Carnaval ACOQ, Graham CH (2013) Environmental correlates of anuran beta diversity in the Brazilian Cerrado. Ecography 36:708–717. https://doi. org/10.1111/j.1600-0587.2012.07374.x

Vasconcelos TS, Santos TG, Haddad CFB et al (2010) Climatic variables and altitude as predictors of anuran species richness and number of reproductive modes in Brazil. J Trop Ecol 26:423–432. https://doi.org/10.1017/S0266467410000167

Vasconcelos TS, Rodríguez MÁ, Hawkins BA (2011) Biogeographic distribution patterns of South American amphibians: a regionalization based on cluster analysis. Natureza Conservação 9:67–72

Vasconcelos TS, Rodríguez MÁ, Hawkins BA (2012) Species distribution modelling as a macro-ecological tool: a case study using New World amphibians. Ecography 35:539–548. https://doi.org/10.1111/j.1600-0587.2011.07050.x

Vasconcelos TS, Prado VHM, da Silva FR et al (2014) Biogeographic distribution patterns and their correlates in the diverse frog fauna of the Atlantic Forest hotspot. PLoS One 9(8):e104130. https://doi.org/10.1371/journal.pone.0104130

Vilela B, Villalobos F (2015) letsR: a new R package for data handling and analysis in macroecology. Methods Ecol Evol 6:1229–1234

Villalobos F, Dobrovolski R, Provete DB et al (2013) Is rich and rare the common share? Describing biodiversity patterns to inform conservation practices for South American anurans. PLoS One 8:e56073. https://doi.org/10.1371/journal.-pone.0056073

Wake DB, Koo MS (2018) Primer: amphibians. Curr Biol 28:R1221–R1242

Wells KD (2007) The ecology and behavior of amphibians. The University of Chicago Press, Chicago

Wiens JJ, Pyron RA, Moen DS (2011) Phylogenetic origin of local-scale diversity patterns and the causes of Amazonian megadiversity. Ecol Lett 14:643–652

Wildlife Conservation Society – WCS, Center for International Earth Science Information Network – CIESIN – Columbia University (2005) Last of the wild project, Version 2, 2005 (LWP-2): Global human footprint dataset (Geographic)

Whittaker RJ, Nogués-Bravo D, Araújo MB (2007) Geographical gradients of species richness: a test of the water-energy conjecture of Hawkins et al. (2003) using European data for five taxa. Glob Ecol Biogeogr 16:76–89

Chapter 2
South American Anurans: Species Diversity and Description Trends Through Time and Space

Abstract Amphibians are especially diverse in the Neotropics and have also one of the highest rates of new species description among terrestrial vertebrates. The first systematic synthesis of South American anurans compiled a list of 1644 species, but there have been no update since the last 19 years. Here, we present a descriptive approach for temporal and spatial patterns of anuran species discoveries in South America, emphasizing trending changes in species description rates and number of researchers authoring a given species description. We recovered 2623 anuran species described in South America between 1758 and mid-2017 from 163 genera and 24 families. There is a high rate of species discovery across time, with at least 10 new descriptions per year in the period examined. Time span to reach multiples of 500 new species has dramatically decreased over time. For instance, it took more than two centuries for the description of 500 species since the first species (1750s), whereas it took about 10–12 years in order to add 500 new anuran species after 1990. Then, the curve of the cumulative anuran species description in South America is far from reaching an asymptote, yet it actually exhibits an exponential shape. Similar historical increase was recorded for the number of authors in papers over time, since descriptions are more collaborative in the last decades. Two major hotspots for new species discovery are depicted herein: (i) the Central and Northern Andes and the adjacent western Amazon (notedly in Ecuador, Peru, and Western Brazil) and (ii) the complex of Brazilian highlands encompassing the Atlantic and Brazilian plateau mountains. These trends are discussed according to singular historical events (including changes in research approaches) and possible explanations for the geographic pattern in species discovery.

Keywords Species discovery · Neotropical anurans · Anuran list · Description rates · Scientometric analysis · Spatiotemporal trends

© Springer Nature Switzerland AG 2019
T. S. Vasconcelos et al., *Biogeographic Patterns of South American Anurans*,
https://doi.org/10.1007/978-3-030-26296-9_2

2.1   Introduction

Estimating how many species inhabit the Earth is continuously intriguing scientists (May 1988; Mora et al. 2011; Costello et al. 2012, 2013). Estimates show that 86% of existing species are to be described (Mora et al. 2011), and rates of species description for several groups are increasing exponentially (Joppa et al. 2011).

Amphibians are especially diverse in the Neotropics and also have one of the highest description rates among terrestrial vertebrates (Jenkins et al. 2013). In the Neotropics, South America has a high relevance due to its evolutionary history context, which then results in high species number and endemism rates (Duellman 1979; Villalobos et al. 2013; Antonelli et al. 2018). The first systematic synthesis of South American amphibians included 1742 species, most of which were made up on anurans (1644 species or 94%) (Duellman 1999). More recently, biodiversity databases have been created during the last decades (e.g., the Global Biodiversity Information Facility (GBIF), www.gbif.org; the International Union for Conservation of Nature's Red List of Threatened Species, www.iucnredlist.org; the Species Link project, http://splink.cria.org.br), some of them specifically developed for amphibians (the portal Amphibian Species of the World, http://research.amnh.org/vz/herpetology/amphibia/index.php; the Amphibia Web portal, http://amphibiaweb.org), which in turn favored the performance of different biogeographical studies on South American anurans (e.g., Vasconcelos et al. 2012; Villalobos et al. 2013). Nonetheless, no anuran systematic update has been performed for anurans in South America since Duellman (1999). In this chapter we present an update on the anuran species list found in South America and perform a descriptive approach for the temporal and spatial patterns of anuran species discoveries. Specifically, we provide information and discuss about rates of species description over time, number of authors involved in these descriptions, and where in the geographical space species have been recently described in South America.

2.2   Material and Methods

2.2.1   Anuran Species List

The anuran species compiled as described in Chap. 1 were initially organized according to the year that each one was described, so they were clustered in decades' time span from the 1750s (the decade of the oldest species) to mid-2017 in order to evaluate temporal trends in the South American anuran descriptions. Therefore, we were able to build a cumulative curve of anuran species description in South America, as well as to calculate the mean number of species described per decade. We also gathered information on the number of the authors each species took for its description in order to analyze a supposed trend in authorships' cooperation across time. The spatial distribution patterns in the anuran description were obtained in order to identify where, in the South America geographical space, new species have been recently discovered. To do so, we took the presence/absence matrix of species occurrence in the South America grid system, as described in Chap. 1, and replaced the code "1" at each grid cell that a given species is present by the year that the respective species was described. Subsequently, we took the mean of each grid cell, so we were able to evaluate the average description year from all species occurring at each grid cell.

2.3   Results and Discussion

We found 2623 anuran species described in South America between 1758 and mid-2017, from 163 genera and 24 families (Table 2.1). At least, two species (*Atopophrynus syntomopus* and *Geobatrachus walkeri*) remain not assigned to any family, yet they are within the superfamily Brachycephaloidea at the time of writing (Table 2.2).

Higher species richness was concentrated within the families Craugastoridae and Hylidae, which, respectively, represent 25% and 20% of anurans in South America (Fig. 2.1). The remarkable species diversity of Craugastoridae is peculiarly related to the megadiverse genus *Pristimantis* that encompass many described and still undescribed species (see Fouquet et al. 2013; Oliveira et al. 2017 and references therein).

An important issue that deserves attention is the dynamic changing of amphibian nomenclature, mainly after the 2000s. The twenty-first century has experienced the integration of different biological disciplines into the traditional taxonomy (e.g., genetics and molecular biology, bioacoustics), so a high number of species that were once allocated in a given family (e.g., *Phyllomedusa* within the family Hylidae; *Craugastor*, assigned as part of *Eleutherodactylus*, within the Leptodactylidae) are now currently allocated in other recently described families (e.g., *Phyllomedusa* within the family Phyllomedusidae and *Craugastor* within the family Craugastoridae). All in all, we are aware that the classification of our anuran compilation may considerably change within the next years/decades (e.g., species can be synonymized, resurrected, or split at any time). Indeed, we agree that this is actually a positive point that reflects more herpetologists engaged in unravelling the evolution of this intriguing vertebrate group in South America.

Historical records of species descriptions from the mid-eighteenth century and the birth of modern taxonomic nomenclature in 1758 (Linnaeus 1758) coincide with the first anuran description in South America. The cumulative curve of the anuran species discoveries in South America has an exponential shape. The first 500 species descriptions took 163 years from the first species described, i.e., until the early twentieth century. However, the time span to reach 500 new described species has dramatically decreased over time; currently, it usually takes 10–12 years to add 500 new South American anurans (Fig. 2.2).

Table 2.1 Species list of anurans occurring in South America according to exhaustive searches at the amphibian species of the World Database (Frost 2017) until mid-2017

Family	Species
Allophrynidae	*Allophryne relicta*
	Allophryne resplendens
	Allophryne ruthveni
Alsodidae	*Alsodes australis*
	Alsodes barrioi
	Alsodes cantillanensis
	Alsodes gargola
	Alsodes hugoi
	Alsodes igneus
	Alsodes kaweshkari
	Alsodes montanus
	Alsodes monticola
	Alsodes nodosus
	Alsodes norae
	Alsodes pehuenche
	Alsodes tumultuosus
	Alsodes valdiviensis
	Alsodes vanzolinii
	Alsodes verrucosus
	Alsodes vittatus
	Eupsophus calcaratus
	Eupsophus contulmoensis
	Eupsophus emiliopugini
	Eupsophus insularis
	Eupsophus migueli
	Eupsophus nahuelbutensis
	Eupsophus queulensis
	Eupsophus roseus
	Eupsophus septentrionalis
	Eupsophus vertebralis
	Limnomedusa macroglossa
Aromobatidae	*Allobates alessandroi*
	Allobates algorei
	Allobates bromelicola
	Allobates brunneus
	Allobates caeruleodactylus
	Allobates caribe
	Allobates cepedai
	Allobates conspicuus
	Allobates crombiei

(continued)

Table 2.1   (continued)

Family	Species
	Allobates femoralis
	Allobates flaviventris
	Allobates fratisenescus
	Allobates fuscellus
	Allobates gasconi
	Allobates goianus
	Allobates granti
	Allobates humilis
	Allobates insperatus
	Allobates juanii
	Allobates kingsburyi
	Allobates mandelorum
	Allobates marchesianus
	Allobates masniger
	Allobates mcdiarmidi
	Allobates melanolaemus
	Allobates myersi
	Allobates nidicola
	Allobates niputidea
	Allobates olfersioides
	Allobates ornatus
	Allobates paleovarzensis
	Allobates picachos
	Allobates pittieri
	Allobates ranoides
	Allobates sanmartini
	Allobates spumaponens
	Allobates subfolionidificans
	Allobates sumtuosus
	Allobates talamancae
	Allobates trilineatus
	Allobates undulatus
	Allobates vanzolinius
	Allobates wayuu
	Allobates zaparo
	Anomaloglossus atopoglossus
	Anomaloglossus ayarzaguenai
	Anomaloglossus baeobatrachus
	Anomaloglossus beebei
	Anomaloglossus breweri
	Anomaloglossus degranvillei
	Anomaloglossus guanayensis

(continued)

Table 2.1 (continued)

Family	Species
	Anomaloglossus kaiei
	Anomaloglossus lacrimosus
	Anomaloglossus murisipanensis
	Anomaloglossus parimae
	Anomaloglossus parkerae
	Anomaloglossus praderioi
	Anomaloglossus roraima
	Anomaloglossus rufulus
	Anomaloglossus shrevei
	Anomaloglossus stepheni
	Anomaloglossus tamacuarensis
	Anomaloglossus tepuyensis
	Anomaloglossus triunfo
	Anomaloglossus wothuja
	Aromobates alboguttatus
	Aromobates capurinensis
	Aromobates duranti
	Aromobates haydeeae
	Aromobates leopardalis
	Aromobates mayorgai
	Aromobates meridensis
	Aromobates molinarii
	Aromobates nocturnus
	Aromobates orostoma
	Aromobates saltuensis
	Aromobates serranus
	Mannophryne caquetio
	Mannophryne collaris
	Mannophryne cordilleriana
	Mannophryne herminae
	Mannophryne lamarcai
	Mannophryne larandina
	Mannophryne leonardoi
	Mannophryne neblina
	Mannophryne oblitterata
	Mannophryne olmonae
	Mannophryne riveroi
	Mannophryne speeri
	Mannophryne trinitatis
	Mannophryne trujillensis
	Mannophryne venezuelensis
	Mannophryne yustizi
	Prostherapis dunni

(continued)

Table 2.1 (continued)

Family	Species
	Rheobates palmatus
	Rheobates pseudopalmatus
Batrachylidae	*Atelognathus ceii*
	Atelognathus jeinimenensis
	Atelognathus nitoi
	Atelognathus patagonicus
	Atelognathus praebasalticus
	Atelognathus reverberii
	Atelognathus salai
	Atelognathus solitarius
	Batrachyla antartandica
	Batrachyla fitzroya
	Batrachyla leptopus
	Batrachyla nibaldoi
	Batrachyla taeniata
	Chaltenobatrachus grandisonae
	Hylorina sylvatica
Brachycephalidae	*Brachycephalus albolineatus*
	Brachycephalus alipioi
	Brachycephalus atelopoide
	Brachycephalus auroguttatus
	Brachycephalus boticario
	Brachycephalus brunneus
	Brachycephalus bufonoides
	Brachycephalus crispus
	Brachycephalus didactylus
	Brachycephalus ephippium
	Brachycephalus ferruginus
	Brachycephalus fuscolineatus
	Brachycephalus garbeanus
	Brachycephalus guarani
	Brachycephalus hermogenesi
	Brachycephalus izecksohni
	Brachycephalus leopardus
	Brachycephalus margariatus
	Brachycephalus mariaeterezae
	Brachycephalus nodoterga
	Brachycephalus olivaceus
	Brachycephalus pernix
	Brachycephalus pitanga
	Brachycephalus pombali
	Brachycephalus pulex

(continued)

Table 2.1 (continued)

Family	Species
	Brachycephalus quiririensis
	Brachycephalus sulfuratus
	Brachycephalus toby
	Brachycephalus tridactylus
	Brachycephalus verrucosus
	Brachycephalus vertebralis
	Ischnocnema abdita
	Ischnocnema bolbodactyla
	Ischnocnema concolor
	Ischnocnema epipeda
	Ischnocnema erythromera
	Ischnocnema gehrti
	Ischnocnema gualteri
	Ischnocnema guentheri
	Ischnocnema henselii
	Ischnocnema hoehnei
	Ischnocnema holti
	Ischnocnema izecksohni
	Ischnocnema juipoca
	Ischnocnema karst
	Ischnocnema lactea
	Ischnocnema manezinho
	Ischnocnema melanopygia
	Ischnocnema nanahallux
	Ischnocnema nasuta
	Ischnocnema nigriventris
	Ischnocnema octavioi
	Ischnocnema oea
	Ischnocnema paranaensis
	Ischnocnema parva
	Ischnocnema penaxavantinho
	Ischnocnema pusilla
	Ischnocnema randorum
	Ischnocnema sambaqui
	Ischnocnema spanios
	Ischnocnema surda
	Ischnocnema venancioi
	Ischnocnema verrucosa
	Ischnocnema vizottoi
Brachycephaloidea	*Atopophrynus syntomopus*
	Geobatrachus walkeri
Bufonidae	*Amazophrynella amazonicola*

(continued)

Table 2.1   (continued)

Family	Species
	Amazophrynella bokermanni
	Amazophrynella javierbustamantei
	Amazophrynella manaos
	Amazophrynella matses
	Amazophrynella minuta
	Amazophrynella vote
	Atelopus andinus
	Atelopus angelito
	Atelopus ardila
	Atelopus arsyecue
	Atelopus arthuri
	Atelopus balios
	Atelopus barbotini
	Atelopus bomolochos
	Atelopus boulengeri
	Atelopus carauta
	Atelopus carbonerensis
	Atelopus carrikeri
	Atelopus certus
	Atelopus chocoensis
	Atelopus chrysocorallus
	Atelopus coynei
	Atelopus cruciger
	Atelopus dimorphus
	Atelopus ebenoides
	Atelopus elegans
	Atelopus epikeisthos
	Atelopus erythropus
	Atelopus eusebianus
	Atelopus eusebiodiazi
	Atelopus exiguus
	Atelopus famelicus
	Atelopus farci
	Atelopus flavescens
	Atelopus franciscus
	Atelopus galactogaster
	Atelopus gigas
	Atelopus glyphus
	Atelopus guanujo
	Atelopus guitarraensis
	Atelopus halihelos
	Atelopus hoogmoedi

(continued)

Table 2.1 (continued)

Family	Species
	Atelopus ignescens
	Atelopus laetissimus
	Atelopus loettersi
	Atelopus longibrachius
	Atelopus longirostris
	Atelopus lozanoi
	Atelopus lynchi
	Atelopus mandingues
	Atelopus marinkellei
	Atelopus mindoensis
	Atelopus minutulus
	Atelopus mittermeieri
	Atelopus monohernandezii
	Atelopus mucubajiensis
	Atelopus muisca
	Atelopus nahumae
	Atelopus nanay
	Atelopus nepiozomus
	Atelopus nicefori
	Atelopus nocturnus
	Atelopus onorei
	Atelopus orcesi
	Atelopus oxapampae
	Atelopus oxyrhynchus
	Atelopus pachydermus
	Atelopus palmatus
	Atelopus pastuso
	Atelopus patazensis
	Atelopus pedimarmoratus
	Atelopus peruensis
	Atelopus petersi
	Atelopus petriruizi
	Atelopus pictiventris
	Atelopus pinangoi
	Atelopus planispina
	Atelopus podocarpus
	Atelopus pulcher
	Atelopus pyrodactylus
	Atelopus quimbaya
	Atelopus reticulatus
	Atelopus sanjosei
	Atelopus seminiferus

(continued)

Table 2.1 (continued)

Family	Species
	Atelopus senex
	Atelopus sernai
	Atelopus simulatus
	Atelopus siranus
	Atelopus sonsonensis
	Atelopus sorianoi
	Atelopus spumarius
	Atelopus spurrelli
	Atelopus subornatus
	Atelopus tamaense
	Atelopus tricolor
	Atelopus vogli
	Atelopus walkeri
	Dendrophryniscus berthalutzae
	Dendrophryniscus brevipollicatus
	Dendrophryniscus carvalhoi
	Dendrophryniscus krausae
	Dendrophryniscus leucomystax
	Dendrophryniscus oreites
	Dendrophryniscus organensis
	Dendrophryniscus proboscideus
	Dendrophryniscus skuki
	Dendrophryniscus stawiarskyi
	Frostius erythrophthalmus
	Frostius pernambucensis
	Incilius coniferus
	Melanophryniscus admirabilis
	Melanophryniscus alipioi
	Melanophryniscus atroluteus
	Melanophryniscus biancae
	Melanophryniscus cambaraensis
	Melanophryniscus cupreuscapularis
	Melanophryniscus devincenzii
	Melanophryniscus dorsalis
	Melanophryniscus estebani
	Melanophryniscus fulvoguttatus
	Melanophryniscus klappenbachi
	Melanophryniscus krauczuki
	Melanophryniscus langonei
	Melanophryniscus macrogranulosus
	Melanophryniscus milanoi
	Melanophryniscus montevidensis

(continued)

Table 2.1 (continued)

Family	Species
	Melanophryniscus moreirae
	Melanophryniscus orejasmirandai
	Melanophryniscus pachyrhynus
	Melanophryniscus paraguayensis
	Melanophryniscus peritus
	Melanophryniscus rubriventris
	Melanophryniscus sanmartini
	Melanophryniscus setiba
	Melanophryniscus simplex
	Melanophryniscus spectabilis
	Melanophryniscus stelzneri
	Melanophryniscus tumifrons
	Melanophryniscus vilavelhensis
	Melanophryniscus xanthostomus
	Metaphryniscus sosai
	Nannophryne apolobambica
	Nannophryne cophotis
	Nannophryne corynetes
	Nannophryne variegata
	Oreophrynella cryptica
	Oreophrynella dendronastes
	Oreophrynella huberi
	Oreophrynella macconnelli
	Oreophrynella nigra
	Oreophrynella quelchii
	Oreophrynella seegobini
	Oreophrynella vasquezi
	Oreophrynella weiassipuensis
	Osornophryne angel
	Osornophryne antisana
	Osornophryne bufoniformis
	Osornophryne cofanorum
	Osornophryne guacamayo
	Osornophryne occidentalis
	Osornophryne percrassa
	Osornophryne puruanta
	Osornophryne simpsoni
	Osornophryne sumacoensis
	Osornophryne talipes
	Rhaebo anderssoni
	Rhaebo andinophrynoides
	Rhaebo atelopoides

(continued)

Table 2.1 (continued)

Family	Species
	Rhaebo blombergi
	Rhaebo caeruleostictus
	Rhaebo colomai
	Rhaebo ecuadorensis
	Rhaebo glaberrimus
	Rhaebo guttatus
	Rhaebo haematiticus
	Rhaebo hypomelas
	Rhaebo lynchi
	Rhaebo nasicus
	Rhaebo olallai
	Rhinella abei
	Rhinella achalensis
	Rhinella achavali
	Rhinella acrolopha
	Rhinella acutirostris
	Rhinella alata
	Rhinella amabilis
	Rhinella amboroensis
	Rhinella arborescandens
	Rhinella arenarum
	Rhinella arequipensis
	Rhinella arunco
	Rhinella atacamensis
	Rhinella azarai
	Rhinella beebei
	Rhinella bergi
	Rhinella bernardoi
	Rhinella casconi
	Rhinella castaneotica
	Rhinella ceratophrys
	Rhinella cerradensis
	Rhinella chavin
	Rhinella chrysophora
	Rhinella cristinae
	Rhinella crucifer
	Rhinella dapsilis
	Rhinella diptycha
	Rhinella dorbignyi
	Rhinella fernandezae
	Rhinella festae
	Rhinella fissipes

(continued)

Table 2.1 (continued)

Family	Species
	Rhinella gallardoi
	Rhinella gildae
	Rhinella gnustae
	Rhinella granulosa
	Rhinella henseli
	Rhinella hoogmoedi
	Rhinella horribilis
	Rhinella humboldti
	Rhinella icterica
	Rhinella inca
	Rhinella inopina
	Rhinella iserni
	Rhinella jimi
	Rhinella justinianoi
	Rhinella leptoscelis
	Rhinella lescurei
	Rhinella limensis
	Rhinella lindae
	Rhinella macrorhina
	Rhinella magnussoni
	Rhinella major
	Rhinella manu
	Rhinella margaritifera
	Rhinella marina
	Rhinella martyi
	Rhinella merianae
	Rhinella mirandaribeiroi
	Rhinella multiverrucosa
	Rhinella nattereri
	Rhinella nesiotes
	Rhinella nicefori
	Rhinella ocellata
	Rhinella ornata
	Rhinella paraguas
	Rhinella paraguayensis
	Rhinella poeppigii
	Rhinella pombali
	Rhinella proboscidea
	Rhinella pygmaea
	Rhinella quechua
	Rhinella roqueana
	Rhinella rostrata

(continued)

Table 2.1 (continued)

Family	Species
	Rhinella rubescens
	Rhinella rubropunctata
	Rhinella ruizi
	Rhinella rumbolli
	Rhinella schneideri
	Rhinella scitula
	Rhinella sclerocephala
	Rhinella sebbeni
	Rhinella spinulosa
	Rhinella stanlaii
	Rhinella sternosignata
	Rhinella tacana
	Rhinella tenrec
	Rhinella truebae
	Rhinella vellardi
	Rhinella veraguensis
	Rhinella veredas
	Rhinella yanachaga
	Rhinella yunga
	Truebella skoptes
	Truebella tothastes
Calyptocephalellidae	*Calyptocephalella gayi*
	Telmatobufo ignotus
	Telmatobufo australis
	Telmatobufo bullocki
	Telmatobufo venustus
Centrolenidae	*Celsiella revocata*
	Celsiella vozmedianoi
	Centrolene acanthidiocephalum
	Centrolene altitudinale
	Centrolene antioquiense
	Centrolene azulae
	Centrolene bacatum
	Centrolene ballux
	Centrolene buckleyi
	Centrolene charapita
	Centrolene condor
	Centrolene daidaleum
	Centrolene geckoideum
	Centrolene gemmatum
	Centrolene guanacarum
	Centrolene heloderma

(continued)

Table 2.1 (continued)

Family	Species
	Centrolene hesperium
	Centrolene huilense
	Centrolene hybrida
	Centrolene lema
	Centrolene lemniscatum
	Centrolene lynchi
	Centrolene medemi
	Centrolene muelleri
	Centrolene notostictum
	Centrolene ocellifera
	Centrolene paezorum
	Centrolene papillahallicum
	Centrolene peristictum
	Centrolene petrophilum
	Centrolene pipilatum
	Centrolene quindianum
	Centrolene robledoi
	Centrolene sabini
	Centrolene sanchezi
	Centrolene savagei
	Centrolene scirtetes
	Centrolene solitaria
	Centrolene venezuelense
	Chimerella corleone
	Chimerella mariaelenae
	Cochranella balionota
	Cochranella croceopodes
	Cochranella duidaeana
	Cochranella erminea
	Cochranella euhystrix
	Cochranella euknemos
	Cochranella geijskesi
	Cochranella granulosa
	Cochranella guayasamini
	Cochranella litoralis
	Cochranella mache
	Cochranella megistra
	Cochranella nola
	Cochranella phryxa
	Cochranella ramirezi
	Cochranella resplendens
	Cochranella riveroi

(continued)

Table 2.1   (continued)

Family	Species
	Cochranella xanthocheridia
	Espadarana andina
	Espadarana audax
	Espadarana callistomma
	Espadarana durrellorum
	Espadarana fernandoi
	Espadarana prosoblepon
	Hyalinobatrachium anachoretus
	Hyalinobatrachium aureoguttatum
	Hyalinobatrachium bergeri
	Hyalinobatrachium cappellei
	Hyalinobatrachium carlesvilai
	Hyalinobatrachium chirripoi
	Hyalinobatrachium colymbiphyllum
	Hyalinobatrachium crurifasciatum
	Hyalinobatrachium duranti
	Hyalinobatrachium eccentricum
	Hyalinobatrachium esmeralda
	Hyalinobatrachium fleischmanni
	Hyalinobatrachium fragile
	Hyalinobatrachium guairarepanense
	Hyalinobatrachium iaspidiense
	Hyalinobatrachium ibama
	Hyalinobatrachium ignioculus
	Hyalinobatrachium kawense
	Hyalinobatrachium mesai
	Hyalinobatrachium mondolfii
	Hyalinobatrachium munozorum
	Hyalinobatrachium nouraguense
	Hyalinobatrachium orientale
	Hyalinobatrachium pallidum
	Hyalinobatrachium pellucidum
	Hyalinobatrachium ruedai
	Hyalinobatrachium tatayoi
	Hyalinobatrachium taylori
	Hyalinobatrachium tricolor
	Hyalinobatrachium valerioi
	Ikakogi tayrona
	Nymphargus anomalus
	Nymphargus armatus
	Nymphargus bejaranoi
	Nymphargus buenaventura

(continued)

Table 2.1 (continued)

Family	Species
	Nymphargus cariticommatus
	Nymphargus chami
	Nymphargus chancas
	Nymphargus cochranae
	Nymphargus cristinae
	Nymphargus garciae
	Nymphargus grandisonae
	Nymphargus griffithsi
	Nymphargus ignotus
	Nymphargus lasgralarias
	Nymphargus laurae
	Nymphargus luminosus
	Nymphargus luteopunctatus
	Nymphargus mariae
	Nymphargus megacheirus
	Nymphargus mixomaculatus
	Nymphargus nephelophila
	Nymphargus ocellatus
	Nymphargus oreonympha
	Nymphargus phenax
	Nymphargus pluvialis
	Nymphargus posadae
	Nymphargus prasinus
	Nymphargus puyoensis
	Nymphargus rosada
	Nymphargus ruizi
	Nymphargus siren
	Nymphargus spilotus
	Nymphargus sucre
	Nymphargus truebae
	Nymphargus vicenteruedai
	Nymphargus wileyi
	Rulyrana adiazeta
	Rulyrana flavopunctata
	Rulyrana mcdiarmidi
	Rulyrana saxiscandens
	Rulyrana spiculata
	Rulyrana susatamai
	Rulyrana tangarana
	Rupirana cardosoi
	Sachatamia albomaculata
	Sachatamia ilex

(continued)

Table 2.1 (continued)

Family	Species
	Sachatamia orejuela
	Sachatamia punctulata
	Teratohyla adenocheira
	Teratohyla amelie
	Teratohyla midas
	Teratohyla pulverata
	Teratohyla spinosa
	Vitreorana antisthenesi
	Vitreorana baliomma
	Vitreorana castroviejoi
	Vitreorana eurygnatha
	Vitreorana gorzulae
	Vitreorana helenae
	Vitreorana parvula
	Vitreorana ritae
	Vitreorana uranoscopa
Ceratophryidae	*Ceratophrys aurita*
	Ceratophrys calcarata
	Ceratophrys cornuta
	Ceratophrys cranwelli
	Ceratophrys joazeirensis
	Ceratophrys ornata
	Ceratophrys stolzmanni
	Ceratophrys testudo
	Chacophrys pierottii
	Lepidobatrachus asper
	Lepidobatrachus laevis
	Lepidobatrachus llanensis
Craugastoridae	*Barycholos pulcher*
	Barycholos ternetzi
	Bryophryne bustamantei
	Bryophryne cophites
	Bryophryne gymnotis
	Bryophryne hanssaueri
	Bryophryne nubilosus
	Bryophryne zonalis
	Ceuthomantis aracamuni
	Ceuthomantis cavernibardus
	Ceuthomantis duellmani
	Craugastor crassidigitus
	Craugastor fitzingeri
	Craugastor longirostris

(continued)

Table 2.1 (continued)

Family	Species
	Craugastor metriosistus
	Craugastor opimus
	Craugastor raniformis
	Dischidodactylus colonnelloi
	Dischidodactylus duidensis
	Euparkerella brasiliensis
	Euparkerella cochranae
	Euparkerella cryptica
	Euparkerella robusta
	Euparkerella tridactyla
	Haddadus aramunha
	Haddadus binotatus
	Haddadus plicifer
	Holoaden bradei
	Holoaden luederwaldti
	Holoaden pholeter
	Holoaden suarezi
	Hypodactylus adercus
	Hypodactylus araiodactylus
	Hypodactylus babax
	Hypodactylus brunneus
	Hypodactylus dolops
	Hypodactylus elassodiscus
	Hypodactylus fallaciosus
	Hypodactylus latens
	Hypodactylus lucida
	Hypodactylus mantipus
	Hypodactylus nigrovittatus
	Hypodactylus peraccai
	Lynchius flavomaculatus
	Lynchius nebulanastes
	Lynchius oblitus
	Lynchius parkeri
	Lynchius simmonsi
	Lynchius tabaconas
	Niceforonia adenobrachia
	Niceforonia columbiana
	Niceforonia nana
	Noblella carrascoicola
	Noblella coloma
	Noblella duellmani
	Noblella heyeri

(continued)

Table 2.1 (continued)

Family	Species
	Noblella lochites
	Noblella lynchi
	Noblella madreselva
	Noblella myrmecoides
	Noblella personina
	Noblella pygmaea
	Noblella ritarasquinae
	Oreobates amarakaeri
	Oreobates ayacucho
	Oreobates barituensis
	Oreobates berdemenos
	Oreobates choristolemma
	Oreobates crepitans
	Oreobates cruralis
	Oreobates discoidalis
	Oreobates gemcare
	Oreobates granulosus
	Oreobates heterodactylus
	Oreobates ibischi
	Oreobates lehri
	Oreobates lundbergi
	Oreobates machiguenga
	Oreobates madidi
	Oreobates pereger
	Oreobates quixensis
	Oreobates remotus
	Oreobates sanctaecrucis
	Oreobates sanderi
	Oreobates saxatilis
	Oreobates yanucu
	Oreobates zongoensis
	Phrynopus auriculatus
	Phrynopus badius
	Phrynopus barthlenae
	Phrynopus bracki
	Phrynopus bufoides
	Phrynopus chaparroi
	Phrynopus curator
	Phrynopus daemon
	Phrynopus dagmarae
	Phrynopus heimorum
	Phrynopus horstpauli

(continued)

Table 2.1 (continued)

Family	Species
	Phrynopus interstinctus
	Phrynopus juninensis
	Phrynopus kauneorum
	Phrynopus kotosh
	Phrynopus miroslawae
	Phrynopus montium
	Phrynopus nicoleae
	Phrynopus oblivius
	Phrynopus paucari
	Phrynopus peruanus
	Phrynopus pesantesi
	Phrynopus tautzorum
	Phrynopus thompsoni
	Phrynopus tribulosus
	Phrynopus valquii
	Phrynopus vestigiatus
	Pristimantis aaptus
	Pristimantis abakapa
	Pristimantis academicus
	Pristimantis acatallelus
	Pristimantis acerus
	Pristimantis achatinus
	Pristimantis achuar
	Pristimantis actinolaimus
	Pristimantis actites
	Pristimantis acuminatus
	Pristimantis acutirostris
	Pristimantis adiastolus
	Pristimantis aemulatus
	Pristimantis affinis
	Pristimantis alalocophus
	Pristimantis albericoi
	Pristimantis albertus
	Pristimantis allpapuyu
	Pristimantis almendariz
	Pristimantis altamazonicus
	Pristimantis altamnis
	Pristimantis ameliae
	Pristimantis amydrotus
	Pristimantis andinognomus
	Pristimantis anemerus
	Pristimantis angustilineatus

(continued)

Table 2.1 (continued)

Family	Species
	Pristimantis aniptopalmatus
	Pristimantis anolirex
	Pristimantis anotis
	Pristimantis apiculatus
	Pristimantis appendiculatus
	Pristimantis aquilonaris
	Pristimantis ardalonychus
	Pristimantis ardyae
	Pristimantis ashaninka
	Pristimantis atrabracus
	Pristimantis atratus
	Pristimantis aurantiguttatus
	Pristimantis aureolineatus
	Pristimantis aureoventris
	Pristimantis auricarens
	Pristimantis avicuporum
	Pristimantis avius
	Pristimantis bacchus
	Pristimantis baiotis
	Pristimantis balionotus
	Pristimantis bambu
	Pristimantis baryecuus
	Pristimantis batrachites
	Pristimantis bearsei
	Pristimantis bellae
	Pristimantis bellator
	Pristimantis bellona
	Pristimantis bernali
	Pristimantis bicantus
	Pristimantis bicolor
	Pristimantis bicumulus
	Pristimantis bipunctatus
	Pristimantis boconoensis
	Pristimantis bogotensis
	Pristimantis boulengeri
	Pristimantis brevifrons
	Pristimantis briceni
	Pristimantis bromeliaceus
	Pristimantis buccinator
	Pristimantis buckleyi
	Pristimantis buenaventura
	Pristimantis bustamante

(continued)

Table 2.1 (continued)

Family	Species
	Pristimantis cabrerai
	Pristimantis cacao
	Pristimantis caeruleonotus
	Pristimantis cajamarcensis
	Pristimantis calcaratus
	Pristimantis calcarulatus
	Pristimantis cantitans
	Pristimantis capitonis
	Pristimantis caprifer
	Pristimantis carlosceroni
	Pristimantis carlossanchezi
	Pristimantis carmelitae
	Pristimantis carranguerorum
	Pristimantis carvalhoi
	Pristimantis caryophyllaceus
	Pristimantis cedros
	Pristimantis celator
	Pristimantis ceuthospilus
	Pristimantis chalceus
	Pristimantis chiastonotus
	Pristimantis chimu
	Pristimantis chloronotus
	Pristimantis chrysops
	Pristimantis citriogaster
	Pristimantis colodactylus
	Pristimantis colomai
	Pristimantis colonensis
	Pristimantis colostichos
	Pristimantis condor
	Pristimantis conservatio
	Pristimantis conspicillatus
	Pristimantis cordovae
	Pristimantis corniger
	Pristimantis coronatus
	Pristimantis corrugatus
	Pristimantis cosnipatae
	Pristimantis cremnobates
	Pristimantis crenunguis
	Pristimantis cristinae
	Pristimantis croceoinguinis
	Pristimantis crucifer
	Pristimantis cruciocularis

(continued)

Table 2.1 (continued)

Family	Species
	Pristimantis cryophilius
	Pristimantis cryptomelas
	Pristimantis cuentasi
	Pristimantis culatensis
	Pristimantis cuneirostris
	Pristimantis curtipes
	Pristimantis danae
	Pristimantis degener
	Pristimantis deinops
	Pristimantis delicatus
	Pristimantis delius
	Pristimantis dendrobatoides
	Pristimantis devillei
	Pristimantis deyi
	Pristimantis diadematus
	Pristimantis diaphonus
	Pristimantis diogenes
	Pristimantis dissimulatus
	Pristimantis divnae
	Pristimantis dorado
	Pristimantis dorsopictus
	Pristimantis duellmani
	Pristimantis duende
	Pristimantis dundeei
	Pristimantis elegans
	Pristimantis enigmaticus
	Pristimantis epacrus
	Pristimantis eremitus
	Pristimantis eriphus
	Pristimantis ernesti
	Pristimantis erythropleura
	Pristimantis esmeraldas
	Pristimantis espedeus
	Pristimantis eugeniae
	Pristimantis eurydactylus
	Pristimantis exoristus
	Pristimantis factiosus
	Pristimantis fallax
	Pristimantis farisorum
	Pristimantis fasciatus
	Pristimantis fenestratus
	Pristimantis festae

(continued)

Table 2.1 (continued)

Family	Species
	Pristimantis fetosus
	Pristimantis flabellidiscus
	Pristimantis flavobracatus
	Pristimantis floridus
	Pristimantis frater
	Pristimantis gaigei
	Pristimantis galdi
	Pristimantis ganonotus
	Pristimantis geminus
	Pristimantis gentryi
	Pristimantis ginesi
	Pristimantis gladiator
	Pristimantis glandulosus
	Pristimantis gracilis
	Pristimantis grandiceps
	Pristimantis grandoculis
	Pristimantis gryllus
	Pristimantis guaiquinimensis
	Pristimantis gualacenio
	Pristimantis gutturalis
	Pristimantis hamiotae
	Pristimantis hampatusami
	Pristimantis hectus
	Pristimantis helvolus
	Pristimantis hernandezi
	Pristimantis hoogmoedi
	Pristimantis huicundo
	Pristimantis hybotragus
	Pristimantis ignicolor
	Pristimantis iiap
	Pristimantis illotus
	Pristimantis imitatrix
	Pristimantis imthurni
	Pristimantis incanus
	Pristimantis incertus
	Pristimantis incomptus
	Pristimantis infraguttatus
	Pristimantis inguinalis
	Pristimantis insignitus
	Pristimantis inusitatus
	Pristimantis ixalus
	Pristimantis jabonensis

(continued)

Table 2.1 (continued)

Family	Species
	Pristimantis jaguensis
	Pristimantis jaimei
	Pristimantis jamescameroni
	Pristimantis jester
	Pristimantis johannesdei
	Pristimantis jorgevelosai
	Pristimantis juanchoi
	Pristimantis jubatus
	Pristimantis karcharias
	Pristimantis kareliae
	Pristimantis katoptroides
	Pristimantis kelephus
	Pristimantis kichwarum
	Pristimantis kirklandi
	Pristimantis koehleri
	Pristimantis kuri
	Pristimantis labiosus
	Pristimantis lacrimosus
	Pristimantis lancinii
	Pristimantis lanthanites
	Pristimantis lasalleorum
	Pristimantis lassoalcalai
	Pristimantis latericius
	Pristimantis laticlavius
	Pristimantis latidiscus
	Pristimantis lemur
	Pristimantis leoni
	Pristimantis leptolophus
	Pristimantis leucopus
	Pristimantis leucorrhinus
	Pristimantis librarius
	Pristimantis lichenoides
	Pristimantis limoncochensis
	Pristimantis lindae
	Pristimantis lirellus
	Pristimantis lividus
	Pristimantis llanganati
	Pristimantis llojsintuta
	Pristimantis longicorpus
	Pristimantis loujosti
	Pristimantis loustes
	Pristimantis lucasi

(continued)

Table 2.1 (continued)

Family	Species
	Pristimantis lucidosignatus
	Pristimantis luscombei
	Pristimantis luteolateralis
	Pristimantis lutitus
	Pristimantis lymani
	Pristimantis lynchi
	Pristimantis lythrodes
	Pristimantis maculosus
	Pristimantis malkini
	Pristimantis marahuaka
	Pristimantis marcoreyesi
	Pristimantis mariaelenae
	Pristimantis marmoratus
	Pristimantis mars
	Pristimantis martiae
	Pristimantis matidiktyo
	Pristimantis mazar
	Pristimantis medemi
	Pristimantis megalops
	Pristimantis melanogaster
	Pristimantis melanoproctus
	Pristimantis memorans
	Pristimantis mendax
	Pristimantis meridionalis
	Pristimantis merostictus
	Pristimantis metabates
	Pristimantis miktos
	Pristimantis mindo
	Pristimantis minimus
	Pristimantis minutulus
	Pristimantis miyatai
	Pristimantis mnionaetes
	Pristimantis modipeplus
	Pristimantis molybrignus
	Pristimantis mondolfii
	Pristimantis moro
	Pristimantis muchimuk
	Pristimantis munozi
	Pristimantis muricatus
	Pristimantis muscosus
	Pristimantis mutabilis
	Pristimantis myersi

(continued)

Table 2.1 (continued)

Family	Species
	Pristimantis myops
	Pristimantis nebulosus
	Pristimantis nephophilus
	Pristimantis nervicus
	Pristimantis nicefori
	Pristimantis nietoi
	Pristimantis nigrogriseus
	Pristimantis nubisilva
	Pristimantis nyctophylax
	Pristimantis obmutescens
	Pristimantis ocellatus
	Pristimantis ockendeni
	Pristimantis ocreatus
	Pristimantis olivaceus
	Pristimantis omeviridis
	Pristimantis onorei
	Pristimantis orcesi
	Pristimantis orcus
	Pristimantis orestes
	Pristimantis ornatissimus
	Pristimantis ornatus
	Pristimantis orpacobates
	Pristimantis orphnolaimus
	Pristimantis ortizi
	Pristimantis padiali
	Pristimantis padrecarlosi
	Pristimantis pahuma
	Pristimantis paisa
	Pristimantis palmeri
	Pristimantis paquishae
	Pristimantis paramerus
	Pristimantis pardalinus
	Pristimantis parectatus
	Pristimantis pariagnomus
	Pristimantis parvillus
	Pristimantis pastazensis
	Pristimantis pataikos
	Pristimantis paulodutrai
	Pristimantis paululus
	Pristimantis pecki
	Pristimantis pedimontanus
	Pristimantis penelopus

(continued)

Table 2.1 (continued)

Family	Species
	Pristimantis peraticus
	Pristimantis percnopterus
	Pristimantis percultus
	Pristimantis permixtus
	Pristimantis peruvianus
	Pristimantis petersi
	Pristimantis petrobardus
	Pristimantis phalaroinguinis
	Pristimantis phalarus
	Pristimantis pharangobates
	Pristimantis philipi
	Pristimantis phoxocephalus
	Pristimantis phragmipleuron
	Pristimantis piceus
	Pristimantis pichincha
	Pristimantis pinchaque
	Pristimantis pinguis
	Pristimantis pirrensis
	Pristimantis platychilus
	Pristimantis platydactylus
	Pristimantis pleurostriatus
	Pristimantis pluvialis
	Pristimantis polemistes
	Pristimantis polychrus
	Pristimantis prolatus
	Pristimantis prometeii
	Pristimantis proserpens
	Pristimantis pruinatus
	Pristimantis pseudoacuminatus
	Pristimantis pteridophilus
	Pristimantis ptochus
	Pristimantis pugnax
	Pristimantis pulchridormientes
	Pristimantis pulvinatus
	Pristimantis punzan
	Pristimantis puruscafeum
	Pristimantis pycnodermis
	Pristimantis pyrrhomerus
	Pristimantis quantus
	Pristimantis quaquaversus
	Pristimantis quicato
	Pristimantis quinquagesimus

(continued)

Table 2.1 (continued)

Family	Species
	Pristimantis racemus
	Pristimantis ramagii
	Pristimantis reclusas
	Pristimantis reichlei
	Pristimantis renjiforum
	Pristimantis repens
	Pristimantis restrepoi
	Pristimantis reticulatus
	Pristimantis rhabdocnemus
	Pristimantis rhabdolaemus
	Pristimantis rhigophilus
	Pristimantis rhodoplichus
	Pristimantis rhodostichus
	Pristimantis ridens
	Pristimantis rivasi
	Pristimantis riveroi
	Pristimantis riveti
	Pristimantis romanorum
	Pristimantis roni
	Pristimantis rosadoi
	Pristimantis roseus
	Pristimantis royi
	Pristimantis rozei
	Pristimantis rubicundus
	Pristimantis ruedai
	Pristimantis rufioculis
	Pristimantis rufoviridis
	Pristimantis ruidus
	Pristimantis ruthveni
	Pristimantis sacharuna
	Pristimantis sagittulus
	Pristimantis salaputium
	Pristimantis saltissimus
	Pristimantis samaipatae
	Pristimantis sanctaemartae
	Pristimantis sanguineus
	Pristimantis sarisarinama
	Pristimantis satagius
	Pristimantis savagei
	Pristimantis schultei
	Pristimantis scitulus
	Pristimantis scoloblepharus

(continued)

Table 2.1 (continued)

Family	Species
	Pristimantis scolodiscus
	Pristimantis scopaeus
	Pristimantis seorsus
	Pristimantis serendipitus
	Pristimantis shrevei
	Pristimantis signifer
	Pristimantis silverstonei
	Pristimantis simonbolivari
	Pristimantis simonsii
	Pristimantis simoteriscus
	Pristimantis simoterus
	Pristimantis siopelus
	Pristimantis sirnigeli
	Pristimantis skydmainos
	Pristimantis sobetes
	Pristimantis spectabilis
	Pristimantis spilogaster
	Pristimantis spinosus
	Pristimantis stenodiscus
	Pristimantis sternothylax
	Pristimantis stictoboubonus
	Pristimantis stictogaster
	Pristimantis stictus
	Pristimantis stipa
	Pristimantis subsigillatus
	Pristimantis suetus
	Pristimantis sulculus
	Pristimantis supernatis
	Pristimantis surdus
	Pristimantis susaguae
	Pristimantis taciturnus
	Pristimantis taeniatus
	Pristimantis tamsitti
	Pristimantis tantanti
	Pristimantis tanyrhynchus
	Pristimantis tayrona
	Pristimantis telefericus
	Pristimantis tenebrionis
	Pristimantis terraebolivaris
	Pristimantis thectopternus
	Pristimantis thyellus
	Pristimantis thymalopsoides

(continued)

Table 2.1 (continued)

Family	Species
	Pristimantis thymelensis
	Pristimantis tinajillas
	Pristimantis tinguichaca
	Pristimantis toftae
	Pristimantis torrenticola
	Pristimantis trachyblepharis
	Pristimantis tribulosus
	Pristimantis truebae
	Pristimantis tubernasus
	Pristimantis tungurahua
	Pristimantis turik
	Pristimantis turumiquirensis
	Pristimantis uisae
	Pristimantis unistrigatus
	Pristimantis urani
	Pristimantis uranobates
	Pristimantis vanadise
	Pristimantis variabilis
	Pristimantis veletis
	Pristimantis ventrigranulosus
	Pristimantis ventriguttatus
	Pristimantis ventrimarmoratus
	Pristimantis verecundus
	Pristimantis versicolor
	Pristimantis vertebralis
	Pristimantis vicarius
	Pristimantis vidua
	Pristimantis viejas
	Pristimantis vilarsi
	Pristimantis vilcabambae
	Pristimantis vinhai
	Pristimantis viridicans
	Pristimantis viridis
	Pristimantis wagteri
	Pristimantis walkeri
	Pristimantis waoranii
	Pristimantis wiensi
	Pristimantis w-nigrum
	Pristimantis xeniolum
	Pristimantis xestus
	Pristimantis xylochobates
	Pristimantis yanezi

(continued)

Table 2.1 (continued)

Family	Species
	Pristimantis yaviensis
	Pristimantis yukpa
	Pristimantis yumbo
	Pristimantis yuruaniensis
	Pristimantis yustizi
	Pristimantis zeuctotylus
	Pristimantis zimmermanae
	Pristimantis zoilae
	Pristimantis zophus
	Psychrophrynella adenopleura
	Psychrophrynella ankohuma
	Psychrophrynella bagrecito
	Psychrophrynella boettgeri
	Psychrophrynella chacaltaya
	Psychrophrynella chaupi
	Psychrophrynella chirihampatu
	Psychrophrynella colla
	Psychrophrynella condoriri
	Psychrophrynella guillei
	Psychrophrynella harveyi
	Psychrophrynella iani
	Psychrophrynella iatamasi
	Psychrophrynella illampu
	Psychrophrynella illimani
	Psychrophrynella kallawaya
	Psychrophrynella katantika
	Psychrophrynella kempffi
	Psychrophrynella melanocheira
	Psychrophrynella pinguis
	Psychrophrynella quimsacruzis
	Psychrophrynella saltator
	Psychrophrynella teqta
	Psychrophrynella usurpator
	Psychrophrynella wettsteini
	Strabomantis anatipes
	Strabomantis anomalus
	Strabomantis biporcatus
	Strabomantis bufoniformis
	Strabomantis cadenai
	Strabomantis cerastes
	Strabomantis cheiroplethus
	Strabomantis cornutus

(continued)

Table 2.1 (continued)

Family	Species
	Strabomantis helonotus
	Strabomantis ingeri
	Strabomantis laticorpus
	Strabomantis necerus
	Strabomantis necopinus
	Strabomantis ruizi
	Strabomantis sulcatus
	Strabomantis zygodactylus
	Tachiramantis douglasi
	Tachiramantis lentiginosus
	Tachiramantis prolixodiscus
	Yunganastes ashkapara
	Yunganastes bisignatus
	Yunganastes fraudator
	Yunganastes mercedesae
	Yunganastes pluvicanorus
Cycloramphidae	*Cycloramphus acangatan*
	Cycloramphus asper
	Cycloramphus bandeirensis
	Cycloramphus bolitoglossus
	Cycloramphus boraceiensis
	Cycloramphus brasiliensis
	Cycloramphus carvalhoi
	Cycloramphus catarinensis
	Cycloramphus cedrensis
	Cycloramphus diringshofeni
	Cycloramphus dubius
	Cycloramphus duseni
	Cycloramphus eleutherodactylus
	Cycloramphus faustoi
	Cycloramphus fuliginosus
	Cycloramphus granulosus
	Cycloramphus izecksohni
	Cycloramphus juimirim
	Cycloramphus lithomimeticus
	Cycloramphus lutzorum
	Cycloramphus migueli
	Cycloramphus mirandaribeiroi
	Cycloramphus ohausi
	Cycloramphus organensis
	Cycloramphus rhyakonastes
	Cycloramphus semipalmatus

(continued)

Table 2.1 (continued)

Family	Species
	Cycloramphus stejnegeri
	Cycloramphus valae
	Thoropa lutzi
	Thoropa megatympanum
	Thoropa miliaris
	Thoropa petropolitana
	Thoropa saxatilis
	Thoropa taophora
	Zachaenus carvalhoi
	Zachaenus parvulus
Dendrobatidae	*Adelphobates castaneoticus*
	Adelphobates galactonotus
	Adelphobates quinquevittatus
	Ameerega andina
	Ameerega bassleri
	Ameerega berohoka
	Ameerega bilinguis
	Ameerega boehmei
	Ameerega boliviana
	Ameerega braccata
	Ameerega cainarachi
	Ameerega erythromos
	Ameerega flavopicta
	Ameerega hahneli
	Ameerega ignipedis
	Ameerega ingeri
	Ameerega macero
	Ameerega parvula
	Ameerega petersi
	Ameerega picta
	Ameerega planipaleae
	Ameerega pongoensis
	Ameerega pulchripecta
	Ameerega rubriventris
	Ameerega silverstonei
	Ameerega simulans
	Ameerega smaragdina
	Ameerega trivittata
	Ameerega yungicola
	Andinobates abditus
	Andinobates altobueyensis
	Andinobates bombetes

(continued)

Table 2.1   (continued)

Family	Species
	Andinobates daleswansoni
	Andinobates dorisswansonae
	Andinobates fulguritus
	Andinobates minutus
	Andinobates opisthomelas
	Andinobates tolimensis
	Andinobates viridis
	Andinobates virolinensis
	Colostethus agilis
	Colostethus alacris
	Colostethus argyrogaster
	Colostethus brachistriatus
	Colostethus dysprosium
	Colostethus fraterdanieli
	Colostethus fugax
	Colostethus furviventris
	Colostethus imbricolus
	Colostethus inguinalis
	Colostethus jacobuspetersi
	Colostethus latinasus
	Colostethus lynchi
	Colostethus mertensi
	Colostethus panamansis
	Colostethus poecilonotus
	Colostethus pratti
	Colostethus ramirezi
	Colostethus ruthveni
	Colostethus thorntoni
	Colostethus ucumari
	Colostethus yaguara
	Dendrobates auratus
	Dendrobates leucomelas
	Dendrobates nubeculosus
	Dendrobates tinctorius
	Dendrobates truncatus
	Epipedobates anthonyi
	Epipedobates boulengeri
	Epipedobates espinosai
	Epipedobates machalilla
	Epipedobates narinensis
	Epipedobates tricolor
	Excidobates captivus

(continued)

Table 2.1 (continued)

Family	Species
	Excidobates mysteriosus
	Hyloxalus abditaurantius
	Hyloxalus aeruginosus
	Hyloxalus anthracinus
	Hyloxalus awa
	Hyloxalus azureiventris
	Hyloxalus betancuri
	Hyloxalus bocagei
	Hyloxalus borjai
	Hyloxalus breviquartus
	Hyloxalus cevallosi
	Hyloxalus chlorocraspedus
	Hyloxalus chocoensis
	Hyloxalus craspedoceps
	Hyloxalus delatorreae
	Hyloxalus edwardsi
	Hyloxalus elachyhistus
	Hyloxalus eleutherodactylus
	Hyloxalus exasperatus
	Hyloxalus excisus
	Hyloxalus faciopunctulatus
	Hyloxalus fallax
	Hyloxalus fascianigrus
	Hyloxalus fuliginosus
	Hyloxalus idiomelus
	Hyloxalus infraguttatus
	Hyloxalus insulatus
	Hyloxalus lehmanni
	Hyloxalus leucophaeus
	Hyloxalus littoralis
	Hyloxalus maculosus
	Hyloxalus maquipucuna
	Hyloxalus marmoreoventris
	Hyloxalus mittermeieri
	Hyloxalus mystax
	Hyloxalus nexipus
	Hyloxalus parcus
	Hyloxalus patitae
	Hyloxalus peculiaris
	Hyloxalus peruvianus
	Hyloxalus pinguis
	Hyloxalus pulchellus

(continued)

Table 2.1   (continued)

Family	Species
	Hyloxalus pulcherrimus
	Hyloxalus pumilus
	Hyloxalus ramosi
	Hyloxalus ruizi
	Hyloxalus saltuarius
	Hyloxalus sauli
	Hyloxalus shuar
	Hyloxalus sordidatus
	Hyloxalus spilotogaster
	Hyloxalus subpunctatus
	Hyloxalus sylvaticus
	Hyloxalus toachi
	Hyloxalus utcubambensis
	Hyloxalus vergeli
	Hyloxalus vertebralis
	Hyloxalus whymperi
	Minyobates steyermarki
	Oophaga histrionica
	Oophaga lehmanni
	Oophaga occultator
	Oophaga sylvatica
	Phyllobates aurotaenia
	Phyllobates bicolor
	Phyllobates terribilis
	Ranitomeya amazonica
	Ranitomeya benedicta
	Ranitomeya duellmani
	Ranitomeya fantastica
	Ranitomeya flavovittata
	Ranitomeya ignea
	Ranitomeya imitator
	Ranitomeya reticulata
	Ranitomeya sirensis
	Ranitomeya summersi
	Ranitomeya uakarii
	Ranitomeya vanzolinii
	Ranitomeya variabilis
	Ranitomeya ventrimaculata
	Silverstoneia erasmios
	Silverstoneia flotator
	Silverstoneia nubicola
Eleutherodactylidae	*Adelophryne adiastola*

(continued)

Table 2.1 (continued)

Family	Species
	Adelophryne baturitensis
	Adelophryne gutturosa
	Adelophryne maranguapensis
	Adelophryne pachydactyla
	Adelophryne patamona
	Diasporus anthrax
	Diasporus gularis
	Diasporus quidditus
	Diasporus tinker
	Eleutherodactylus bilineatus
	Eleutherodactylus johnstonei
	Phyzelaphryne miriamae
Hemiphractidae	*Cryptobatrachus boulengeri*
	Cryptobatrachus fuhrmanni
	Flectonotus fitzgeraldi
	Flectonotus pygmaeus
	Fritziana fissilis
	Fritziana goeldii
	Fritziana ohausi
	Gastrotheca abdita
	Gastrotheca aguaruna
	Gastrotheca albolineata
	Gastrotheca andaquiensis
	Gastrotheca angustifrons
	Gastrotheca antomia
	Gastrotheca antoniiochoai
	Gastrotheca aratia
	Gastrotheca argenteovirens
	Gastrotheca atympana
	Gastrotheca aureomaculata
	Gastrotheca bufona
	Gastrotheca carinaceps
	Gastrotheca christiani
	Gastrotheca chrysosticta
	Gastrotheca cornuta
	Gastrotheca dendronastes
	Gastrotheca dunni
	Gastrotheca ernestoi
	Gastrotheca espeletia
	Gastrotheca excubitor
	Gastrotheca fissipes
	Gastrotheca flamma

(continued)

Table 2.1 (continued)

Family	Species
	Gastrotheca fulvorufa
	Gastrotheca galeata
	Gastrotheca gracilis
	Gastrotheca griswoldi
	Gastrotheca guentheri
	Gastrotheca helenae
	Gastrotheca lateonota
	Gastrotheca lauzuricae
	Gastrotheca litonedis
	Gastrotheca longipes
	Gastrotheca marsupiata
	Gastrotheca microdiscus
	Gastrotheca monticola
	Gastrotheca nicefori
	Gastrotheca ochoai
	Gastrotheca orophylax
	Gastrotheca ossilaginis
	Gastrotheca ovifera
	Gastrotheca pacchamama
	Gastrotheca peruana
	Gastrotheca phalarosa
	Gastrotheca piperata
	Gastrotheca plumbea
	Gastrotheca pseustes
	Gastrotheca psychrophila
	Gastrotheca rebeccae
	Gastrotheca riobambae
	Gastrotheca ruizi
	Gastrotheca splendens
	Gastrotheca stictopleura
	Gastrotheca testudinea
	Gastrotheca trachyceps
	Gastrotheca walkeri
	Gastrotheca weinlandii
	Gastrotheca williamsoni
	Gastrotheca zeugocystis
	Hemiphractus bubalus
	Hemiphractus fasciatus
	Hemiphractus helioi
	Hemiphractus johnsoni
	Hemiphractus proboscideus
	Hemiphractus scutatus

(continued)

Table 2.1 (continued)

Family	Species
	Stefania ackawaio
	Stefania ayangannae
	Stefania breweri
	Stefania coxi
	Stefania evansi
	Stefania ginesi
	Stefania goini
	Stefania marahuaquensis
	Stefania oculosa
	Stefania percristata
	Stefania riae
	Stefania riveroi
	Stefania roraimae
	Stefania satelles
	Stefania scalae
	Stefania schuberti
	Stefania tamacuarina
	Stefania woodleyi
Hylidae	*Aparasphenodon arapapa*
	Aparasphenodon bokermanni
	Aparasphenodon brunoi
	Aparasphenodon pomba
	Aparasphenodon venezolanus
	Aplastodiscus albofrenatus
	Aplastodiscus albosignatus
	Aplastodiscus arildae
	Aplastodiscus cavicola
	Aplastodiscus cochranae
	Aplastodiscus ehrhardti
	Aplastodiscus eugenioi
	Aplastodiscus flumineus
	Aplastodiscus ibirapitanga
	Aplastodiscus leucopygius
	Aplastodiscus lutzorum
	Aplastodiscus musicus
	Aplastodiscus perviridis
	Aplastodiscus sibilatus
	Aplastodiscus weygoldti
	Argenteohyla siemersi
	Boana aguilari
	Boana albomarginata
	Boana albonigra

(continued)

Table 2.1 (continued)

Family	Species
	Boana albopunctata
	Boana alemani
	Boana alfaroi
	Boana almendarizae
	Boana atlantica
	Boana balzani
	Boana bandeirantes
	Boana beckeri
	Boana benitezi
	Boana bischoffi
	Boana boans
	Boana botumirim
	Boana buriti
	Boana caipora
	Boana caingua
	Boana calcarata
	Boana callipleura
	Boana cambui
	Boana cinerascens
	Boana cipoensis
	Boana cordobae
	Boana crepitans
	Boana curupi
	Boana cymbalum
	Boana dentei
	Boana diabolica
	Boana ericae
	Boana exastis
	Boana faber
	Boana fasciata
	Boana freicanecae
	Boana geographica
	Boana gladiator
	Boana goiana
	Boana guentheri
	Boana hobbsi
	Boana hutchinsi
	Boana jaguariaivensis
	Boana jimenezi
	Boana joaquini
	Boana lanciformis
	Boana latistriata

(continued)

Table 2.1 (continued)

Family	Species
	Boana lemai
	Boana leptolineata
	Boana leucocheila
	Boana liliae
	Boana lundii
	Boana maculateralis
	Boana marginata
	Boana marianitae
	Boana melanopleura
	Boana microderma
	Boana multifasciata
	Boana nympha
	Boana ornatissima
	Boana palaestes
	Boana paranaiba
	Boana pardalis
	Boana pellucens
	Boana phaeopleura
	Boana picturata
	Boana poaju
	Boana polytaenia
	Boana pombali
	Boana prasina
	Boana pugnax
	Boana pulchella
	Boana pulidoi
	Boana punctata
	Boana raniceps
	Boana rhythmica
	Boana riojana
	Boana roraima
	Boana rosenbergi
	Boana rubracyla
	Boana rufitela
	Boana secedens
	Boana semiguttata
	Boana semilineata
	Boana sibleszi
	Boana steinbachi
	Boana stellae
	Boana stenocephala
	Boana tepuiana

(continued)

Table 2.1 (continued)

Family	Species
	Boana tetete
	Boana varelae
	Boana wavrini
	Boana xerophylla
	Bokermannohyla ahenea
	Bokermannohyla alvarengai
	Bokermannohyla astartea
	Bokermannohyla capra
	Bokermannohyla caramaschii
	Bokermannohyla carvalhoi
	Bokermannohyla circumdata
	Bokermannohyla claresignata
	Bokermannohyla clepsydra
	Bokermannohyla diamantina
	Bokermannohyla flavopicta
	Bokermannohyla gouveai
	Bokermannohyla hylax
	Bokermannohyla ibitiguara
	Bokermannohyla ibitipoca
	Bokermannohyla itapoty
	Bokermannohyla izecksohni
	Bokermannohyla juiju
	Bokermannohyla langei
	Bokermannohyla lucianae
	Bokermannohyla luctuosa
	Bokermannohyla martinsi
	Bokermannohyla nanuzae
	Bokermannohyla napolii
	Bokermannohyla oxente
	Bokermannohyla pseudopseudis
	Bokermannohyla ravida
	Bokermannohyla sagarana
	Bokermannohyla sapiranga
	Bokermannohyla saxicola
	Bokermannohyla sazimai
	Bokermannohyla vulcaniae
	Colomascirtus antioquia
	Colomascirtus condor
	Colomascirtus criptico
	Colomascirtus princecharlesi
	Colomascirtus tigrinus
	Corythomantis galeata

(continued)

Table 2.1 (continued)

Family	Species
	Corythomantis greeningi
	Dendropsophus acreanus
	Dendropsophus amicorum
	Dendropsophus anataliasiasi
	Dendropsophus anceps
	Dendropsophus aperomeus
	Dendropsophus araguaya
	Dendropsophus arndti
	Dendropsophus battersbyi
	Dendropsophus berthalutzae
	Dendropsophus bifurcus
	Dendropsophus bipunctatus
	Dendropsophus bogerti
	Dendropsophus bokermanni
	Dendropsophus branneri
	Dendropsophus brevifrons
	Dendropsophus bromeliaceus
	Dendropsophus cachimbo
	Dendropsophus carnifex
	Dendropsophus cerradensis
	Dendropsophus coffeus
	Dendropsophus columbianus
	Dendropsophus counani
	Dendropsophus cruzi
	Dendropsophus decipiens
	Dendropsophus delarivai
	Dendropsophus dutrai
	Dendropsophus ebraccatus
	Dendropsophus elegans
	Dendropsophus elianeae
	Dendropsophus frosti
	Dendropsophus garagoensis
	Dendropsophus gaucheri
	Dendropsophus giesleri
	Dendropsophus grandisonae
	Dendropsophus gryllatus
	Dendropsophus haddadi
	Dendropsophus haraldschultzi
	Dendropsophus jimi
	Dendropsophus joannae
	Dendropsophus juliani
	Dendropsophus koechlini

(continued)

Table 2.1 (continued)

Family	Species
	Dendropsophus leali
	Dendropsophus leucophyllatus
	Dendropsophus limai
	Dendropsophus luddeckei
	Dendropsophus luteoocellatus
	Dendropsophus manonegra
	Dendropsophus mapinguari
	Dendropsophus marmoratus
	Dendropsophus mathiassoni
	Dendropsophus melanargyreus
	Dendropsophus meridensis
	Dendropsophus meridianus
	Dendropsophus microcephalus
	Dendropsophus microps
	Dendropsophus minimus
	Dendropsophus minusculus
	Dendropsophus minutus
	Dendropsophus miyatai
	Dendropsophus molitor
	Dendropsophus nahdereri
	Dendropsophus nanus
	Dendropsophus norandicus
	Dendropsophus novaisi
	Dendropsophus oliveirai
	Dendropsophus ozzyi
	Dendropsophus padreluna
	Dendropsophus parviceps
	Dendropsophus pauiniensis
	Dendropsophus phlebodes
	Dendropsophus praestans
	Dendropsophus pseudomeridianus
	Dendropsophus reichlei
	Dendropsophus reticulatus
	Dendropsophus rhea
	Dendropsophus rhodopeplus
	Dendropsophus riveroi
	Dendropsophus rossalleni
	Dendropsophus rubicundulus
	Dendropsophus ruschii
	Dendropsophus salli
	Dendropsophus sanborni
	Dendropsophus sarayacuensis

(continued)

Table 2.1 (continued)

Family	Species
	Dendropsophus schubarti
	Dendropsophus seniculus
	Dendropsophus shiwiarum
	Dendropsophus soaresi
	Dendropsophus stingi
	Dendropsophus studerae
	Dendropsophus subocularis
	Dendropsophus timbeba
	Dendropsophus tintinnabulum
	Dendropsophus triangulum
	Dendropsophus tritaeniatus
	Dendropsophus virolinensis
	Dendropsophus vraemi
	Dendropsophus walfordi
	Dendropsophus werneri
	Dendropsophus xapuriensis
	Dendropsophus yaracuyanus
	Dryaderces inframaculata
	Dryaderces pearsoni
	Ecnomiohyla miliaria
	Ecnomiohyla phantasmagoria
	Ecnomiohyla thysanota
	Hyloscirtus albopunctulatus
	Hyloscirtus alytolylax
	Hyloscirtus armatus
	Hyloscirtus bogotensis
	Hyloscirtus callipeza
	Hyloscirtus caucanus
	Hyloscirtus charazani
	Hyloscirtus chlorosteus
	Hyloscirtus colymba
	Hyloscirtus denticulentus
	Hyloscirtus diabolus
	Hyloscirtus estevesi
	Hyloscirtus jahni
	Hyloscirtus larinopygion
	Hyloscirtus lascinius
	Hyloscirtus lindae
	Hyloscirtus lynchi
	Hyloscirtus mashpi
	Hyloscirtus pacha
	Hyloscirtus palmeri

(continued)

Table 2.1   (continued)

Family	Species
	Hyloscirtus pantostictus
	Hyloscirtus phyllognathus
	Hyloscirtus piceigularis
	Hyloscirtus platydactylus
	Hyloscirtus psarolaimus
	Hyloscirtus ptychodactylus
	Hyloscirtus sarampiona
	Hyloscirtus simmonsi
	Hyloscirtus staufferorum
	Hyloscirtus tapichalaca
	Hyloscirtus torrenticola
	Itapotihyla langsdorffii
	Julianus pinimus
	Julianus uruguayus
	Lysapsus bolivianus
	Lysapsus caraya
	Lysapsus laevis
	Lysapsus limellum
	Myersiohyla aromatica
	Myersiohyla chamaeleo
	Myersiohyla inparquesi
	Myersiohyla kanaima
	Myersiohyla loveridgei
	Myersiohyla neblinaria
	Nyctimantis rugiceps
	Ololygon agilis
	Ololygon albicans
	Ololygon alcatraz
	Ololygon angrensis
	Ololygon arduous
	Ololygon argyreornata
	Ololygon ariadne
	Ololygon aromothyella
	Ololygon atrata
	Ololygon belloni
	Ololygon berthae
	Ololygon brieni
	Ololygon caissara
	Ololygon canastrensis
	Ololygon carnevallii
	Ololygon catharinae
	Ololygon centralis

(continued)

Table 2.1 (continued)

Family	Species
	Ololygon cosenzai
	Ololygon faivovichi
	Ololygon flavoguttata
	Ololygon heyeri
	Ololygon hiemalis
	Ololygon humilis
	Ololygon insperata
	Ololygon jureia
	Ololygon kautskyi
	Ololygon littoralis
	Ololygon littoreus
	Ololygon longilinea
	Ololygon luizotavioi
	Ololygon machadoi
	Ololygon melanodactylus
	Ololygon melloi
	Ololygon muriciensis
	Ololygon obtriangulata
	Ololygon peixotoi
	Ololygon perpusilla
	Ololygon pombali
	Ololygon ranki
	Ololygon rizibilis
	Ololygon skaios
	Ololygon skuki
	Ololygon strigilata
	Ololygon trapicheiroi
	Ololygon tripui
	Ololygon tupinamba
	Ololygon v-signata
	Osteocephalus alboguttatus
	Osteocephalus buckleyi
	Osteocephalus cabrerai
	Osteocephalus camufatus
	Osteocephalus cannatellai
	Osteocephalus carri
	Osteocephalus castaneicola
	Osteocephalus deridens
	Osteocephalus duellmani
	Osteocephalus mimeticus
	Osteocephalus festae
	Osteocephalus fuscifacies

(continued)

Table 2.1 (continued)

Family	Species
	Osteocephalus helenae
	Osteocephalus heyeri
	Osteocephalus leoniae
	Osteocephalus leprieurii
	Osteocephalus mutabor
	Osteocephalus oophagus
	Osteocephalus planiceps
	Osteocephalus subtilis
	Osteocephalus taurinus
	Osteocephalus verruciger
	Osteocephalus vilarsi
	Osteocephalus yasuni
	Phyllodytes acuminatus
	Phyllodytes brevirostris
	Phyllodytes edelmoi
	Phyllodytes gyrinaethes
	Phyllodytes kautskyi
	Phyllodytes luteolus
	Phyllodytes maculosus
	Phyllodytes megatympanum
	Phyllodytes melanomystax
	Phyllodytes punctatus
	Phyllodytes tuberculosus
	Phyllodytes wuchereri
	Phyllomedusa araguari
	Phyllomedusa bahiana
	Phyllomedusa bicolor
	Phyllomedusa boliviana
	Phyllomedusa burmeisteri
	Phyllomedusa camba
	Phyllomedusa coelestis
	Phyllomedusa distincta
	Phyllomedusa iheringii
	Phyllomedusa neildi
	Phyllomedusa sauvagii
	Phyllomedusa tarsius
	Phyllomedusa tetraploidea
	Phyllomedusa trinitatis
	Phyllomedusa vaillantii
	Phyllomedusa venusta
	Phytotriades auratus
	Pseudis bolbodactyla

(continued)

Table 2.1 (continued)

Family	Species
	Pseudis cardosoi
	Pseudis fusca
	Pseudis minuta
	Pseudis paradoxa
	Pseudis platensis
	Pseudis tocantins
	Scarthyla goinorum
	Scarthyla vigilans
	Scinax acuminatus
	Scinax alter
	Scinax auratus
	Scinax baumgardneri
	Scinax blairi
	Scinax boesemani
	Scinax boulengeri
	Scinax cabralensis
	Scinax caldarum
	Scinax camposseabrai
	Scinax cardosoi
	Scinax castroviejoi
	Scinax chiquitanus
	Scinax constrictus
	Scinax cretatus
	Scinax crospedospilus
	Scinax cruentommus
	Scinax curicica
	Scinax cuspidatus
	Scinax danae
	Scinax dolloi
	Scinax duartei
	Scinax elaeochroa
	Scinax eurydice
	Scinax exiguus
	Scinax funereus
	Scinax fuscomarginatus
	Scinax fuscovarius
	Scinax garbei
	Scinax granulatus
	Scinax haddadorum
	Scinax hayii
	Scinax ictericus
	Scinax imbegue

(continued)

Table 2.1   (continued)

Family	Species
	Scinax iquitorum
	Scinax jolyi
	Scinax juncae
	Scinax karenanneae
	Scinax kennedyi
	Scinax lindsayi
	Scinax madeirae
	Scinax manriquei
	Scinax maracaya
	Scinax montivagus
	Scinax nasicus
	Scinax nebulosus
	Scinax oreites
	Scinax pachycrus
	Scinax pedromedinae
	Scinax perereca
	Scinax proboscideus
	Scinax quinquefasciatus
	Scinax rogerioi
	Scinax rossaferesae
	Scinax rostratus
	Scinax ruber
	Scinax rupestris
	Scinax sateremawe
	Scinax similis
	Scinax squalirostris
	Scinax sugillatus
	Scinax tigrinus
	Scinax tymbamirim
	Scinax villasboasi
	Scinax wandae
	Scinax x-signatus
	Smilisca phaeota
	Smilisca sila
	Smilisca sordida
	Sphaenorhynchus botocudo
	Sphaenorhynchus bromelicola
	Sphaenorhynchus canga
	Sphaenorhynchus caramaschii
	Sphaenorhynchus carneus
	Sphaenorhynchus dorisae
	Sphaenorhynchus lacteus

(continued)

Table 2.1 (continued)

Family	Species
	Sphaenorhynchus mirim
	Sphaenorhynchus orophilus
	Sphaenorhynchus palustris
	Sphaenorhynchus pauloalvini
	Sphaenorhynchus planicola
	Sphaenorhynchus prasinus
	Sphaenorhynchus surdus
	Tepuihyla aecii
	Tepuihyla edelcae
	Tepuihyla exophthalma
	Tepuihyla luteolabris
	Tepuihyla obscura
	Tepuihyla rodriguezi
	Tepuihyla shushupe
	Tepuihyla tuberculosa
	Tepuihyla warreni
	Trachycephalus atlas
	Trachycephalus coriaceus
	Trachycephalus cunauaru
	Trachycephalus dibernardoi
	Trachycephalus hadroceps
	Trachycephalus helioi
	Trachycephalus imitatrix
	Trachycephalus jordani
	Trachycephalus lepidus
	Trachycephalus macrotis
	Trachycephalus mambaiensis
	Trachycephalus mesophaeus
	Trachycephalus nigromaculatus
	Trachycephalus quadrangulum
	Trachycephalus resinifictrix
	Trachycephalus typhonius
	Xenohyla eugenioi
	Xenohyla truncata
Hylodidae	*Crossodactylus aeneus*
	Crossodactylus bokermanni
	Crossodactylus caramaschii
	Crossodactylus cyclospinus
	Crossodactylus dantei
	Crossodactylus dispar
	Crossodactylus fransciscanus
	Crossodactylus gaudichaudii

(continued)

Table 2.1 (continued)

Family	Species
	Crossodactylus grandis
	Crossodactylus lutzorum
	Crossodactylus schmidti
	Crossodactylus timbuhy
	Crossodactylus trachystomus
	Crossodactylus werneri
	Hylodes amnicola
	Hylodes asper
	Hylodes babax
	Hylodes cardosoi
	Hylodes charadranaetes
	Hylodes dactylocinus
	Hylodes fredi
	Hylodes glaber
	Hylodes heyeri
	Hylodes japi
	Hylodes lateristrigatus
	Hylodes magalhaesi
	Hylodes meridionalis
	Hylodes mertensi
	Hylodes nasus
	Hylodes ornatus
	Hylodes otavioi
	Hylodes perere
	Hylodes perplicatus
	Hylodes phyllodes
	Hylodes pipilans
	Hylodes regius
	Hylodes sazimai
	Hylodes uai
	Hylodes vanzolinii
	Megaelosia apuana
	Megaelosia bocainensis
	Megaelosia boticariana
	Megaelosia goeldii
	Megaelosia jordanensis
	Megaelosia lutzae
	Megaelosia massarti
Leptodactylidae	*Adenomera ajurauna*
	Adenomera andreae
	Adenomera araucaria
	Adenomera bokermanni

(continued)

Table 2.1 (continued)

Family	Species
	Adenomera coca
	Adenomera cotuba
	Adenomera diptyx
	Adenomera engelsi
	Adenomera heyeri
	Adenomera hylaedactyla
	Adenomera juikitam
	Adenomera lutzi
	Adenomera marmorata
	Adenomera martinezi
	Adenomera nana
	Adenomera saci
	Adenomera simonstuarti
	Adenomera thomei
	Crossodactylodes bokermanni
	Crossodactylodes izecksohni
	Crossodactylodes pintoi
	Edalorhina nasuta
	Edalorhina perezi
	Engystomops coloradorum
	Engystomops freibergi
	Engystomops guayaco
	Engystomops montubio
	Engystomops petersi
	Engystomops pustulatus
	Engystomops pustulosus
	Engystomops puyango
	Engystomops randi
	Hydrolaetare caparu
	Hydrolaetare dantasi
	Hydrolaetare schmidti
	Leptodactylus bolivianus
	Leptodactylus bufonius
	Leptodactylus caatingae
	Leptodactylus camaquara
	Leptodactylus chaquensis
	Leptodactylus colombiensis
	Leptodactylus cunicularius
	Leptodactylus cupreus
	Leptodactylus didymus
	Leptodactylus diedrus
	Leptodactylus discodactylus

(continued)

Table 2.1   (continued)

Family	Species
	Leptodactylus elenae
	Leptodactylus flavopictus
	Leptodactylus fragilis
	Leptodactylus furnarius
	Leptodactylus fuscus
	Leptodactylus gracilis
	Leptodactylus griseigularis
	Leptodactylus guianensis
	Leptodactylus hylodes
	Leptodactylus jolyi
	Leptodactylus knudseni
	Leptodactylus labrosus
	Leptodactylus labyrinthicus
	Leptodactylus laticeps
	Leptodactylus latinasus
	Leptodactylus latrans
	Leptodactylus lauramiriamae
	Leptodactylus leptodactyloides
	Leptodactylus lithonaetes
	Leptodactylus longirostris
	Leptodactylus macrosternum
	Leptodactylus magistris
	Leptodactylus marambaiae
	Leptodactylus melanonotus
	Leptodactylus myersi
	Leptodactylus mystaceus
	Leptodactylus mystacinus
	Leptodactylus natalensis
	Leptodactylus nesiotus
	Leptodactylus notoaktites
	Leptodactylus oreomantis
	Leptodactylus pallidirostris
	Leptodactylus paraensis
	Leptodactylus pascoensis
	Leptodactylus pentadactylus
	Leptodactylus peritoaktites
	Leptodactylus petersii
	Leptodactylus plaumanni
	Leptodactylus podicipinus
	Leptodactylus poecilochilus
	Leptodactylus pustulatus
	Leptodactylus rhodomerus

(continued)

Table 2.1 (continued)

Family	Species
	Leptodactylus rhodomystax
	Leptodactylus rhodonotus
	Leptodactylus rhodostima
	Leptodactylus riveroi
	Leptodactylus rugosus
	Leptodactylus sabanensis
	Leptodactylus savagei
	Leptodactylus sertanejo
	Leptodactylus silvanimbus
	Leptodactylus spixi
	Leptodactylus stenodema
	Leptodactylus syphax
	Leptodactylus tapiti
	Leptodactylus troglodytes
	Leptodactylus turimiquensis
	Leptodactylus validus
	Leptodactylus vastus
	Leptodactylus ventrimaculatus
	Leptodactylus viridis
	Leptodactylus wagneri
	Paratelmatobius cardosoi
	Paratelmatobius gaigeae
	Paratelmatobius lutzii
	Paratelmatobius mantiqueira
	Paratelmatobius poecilogaster
	Paratelmatobius yepiranga
	Physalaemus aguirrei
	Physalaemus albifrons
	Physalaemus albonotatus
	Physalaemus angrensis
	Physalaemus atim
	Physalaemus atlanticus
	Physalaemus barrioi
	Physalaemus biligonigerus
	Physalaemus bokermanni
	Physalaemus caete
	Physalaemus camacan
	Physalaemus centralis
	Physalaemus cicada
	Physalaemus crombiei
	Physalaemus cuqui
	Physalaemus cuvieri

(continued)

Table 2.1 (continued)

Family	Species
	Physalaemus deimaticus
	Physalaemus ephippifer
	Physalaemus erikae
	Physalaemus erythros
	Physalaemus evangelistai
	Physalaemus feioi
	Physalaemus fernandezae
	Physalaemus fischeri
	Physalaemus gracilis
	Physalaemus henselii
	Physalaemus insperatus
	Physalaemus irroratus
	Physalaemus jordanensis
	Physalaemus kroyeri
	Physalaemus lateristriga
	Physalaemus lisei
	Physalaemus maculiventris
	Physalaemus marmoratus
	Physalaemus maximus
	Physalaemus moreirae
	Physalaemus nanus
	Physalaemus nattereri
	Physalaemus obtectus
	Physalaemus olfersii
	Physalaemus orophilus
	Physalaemus riograndensis
	Physalaemus rupestris
	Physalaemus santafecinus
	Physalaemus signifer
	Physalaemus soaresi
	Physalaemus spiniger
	Pleurodema alium
	Pleurodema bibroni
	Pleurodema borellii
	Pleurodema brachyops
	Pleurodema bufoninum
	Pleurodema cinereum
	Pleurodema cordobae
	Pleurodema diplolister
	Pleurodema fuscomaculatum
	Pleurodema guayapae
	Pleurodema kriegi

(continued)

Table 2.1 (continued)

Family	Species
	Pleurodema marmoratum
	Pleurodema nebulosum
	Pleurodema somuncurense
	Pleurodema thaul
	Pleurodema tucumanum
	Pseudopaludicola ameghini
	Pseudopaludicola atragula
	Pseudopaludicola boliviana
	Pseudopaludicola canga
	Pseudopaludicola ceratophyes
	Pseudopaludicola facurae
	Pseudopaludicola falcipes
	Pseudopaludicola giarettai
	Pseudopaludicola hyleaustralis
	Pseudopaludicola ibisoroca
	Pseudopaludicola jaredi
	Pseudopaludicola llanera
	Pseudopaludicola mineira
	Pseudopaludicola mirandae
	Pseudopaludicola motorzinho
	Pseudopaludicola murundu
	Pseudopaludicola mystacalis
	Pseudopaludicola parnaiba
	Pseudopaludicola pocoto
	Pseudopaludicola pusilla
	Pseudopaludicola saltica
	Pseudopaludicola ternetzi
	Scythrophrys sawayae
Microhylidae	*Adelastes hylonomos*
	Arcovomer passarellii
	Chiasmocleis alagoana
	Chiasmocleis albopunctata
	Chiasmocleis altomontana
	Chiasmocleis anatipes
	Chiasmocleis antenori
	Chiasmocleis atlantica
	Chiasmocleis avilapiresae
	Chiasmocleis bassleri
	Chiasmocleis capixaba
	Chiasmocleis carvalhoi
	Chiasmocleis centralis
	Chiasmocleis cordeiroi

(continued)

Table 2.1 (continued)

Family	Species
	Chiasmocleis crucis
	Chiasmocleis devriesi
	Chiasmocleis gnoma
	Chiasmocleis haddadi
	Chiasmocleis hudsoni
	Chiasmocleis jimi
	Chiasmocleis lacrimae
	Chiasmocleis leucosticta
	Chiasmocleis magnova
	Chiasmocleis mantiqueira
	Chiasmocleis mehelyi
	Chiasmocleis migueli
	Chiasmocleis papachibe
	Chiasmocleis quilombola
	Chiasmocleis royi
	Chiasmocleis sapiranga
	Chiasmocleis schubarti
	Chiasmocleis shudikarensis
	Chiasmocleis supercilialba
	Chiasmocleis tridactyla
	Chiasmocleis ventrimaculata
	Chiasmocleis veracruz
	Ctenophryne aequatorialis
	Ctenophryne aterrima
	Ctenophryne barbatula
	Ctenophryne carpish
	Ctenophryne geayi
	Ctenophryne minor
	Dasypops schirchi
	Dermatonotus muelleri
	Elachistocleis bicolor
	Elachistocleis bumbameuboi
	Elachistocleis carvalhoi
	Elachistocleis cesarii
	Elachistocleis erythrogaster
	Elachistocleis haroi
	Elachistocleis helianneae
	Elachistocleis magnus
	Elachistocleis matogrosso
	Elachistocleis muiraquitan
	Elachistocleis ovalis
	Elachistocleis panamensis

(continued)

Table 2.1 (continued)

Family	Species
	Elachistocleis pearsei
	Elachistocleis piauiensis
	Elachistocleis skotogaster
	Elachistocleis surinamensis
	Elachistocleis surumu
	Hamptophryne alios
	Hamptophryne boliviana
	Myersiella microps
	Otophryne pyburni
	Otophryne robusta
	Otophryne steyermarki
	Stereocyclops histrio
	Stereocyclops incrassatus
	Stereocyclops palmipes
	Stereocyclops parkeri
	Synapturanus mirandaribeiroi
	Synapturanus rabus
	Synapturanus salseri
Odontophrynidae	*Macrogenioglottus alipioi*
	Odontophrynus achalensis
	Odontophrynus americanus
	Odontophrynus barrioi
	Odontophrynus carvalhoi
	Odontophrynus cordobae
	Odontophrynus cultripes
	Odontophrynus lavillai
	Odontophrynus maisuma
	Odontophrynus monachus
	Odontophrynus occidentalis
	Odontophrynus salvatori
	Proceratophrys appendiculata
	Proceratophrys aridus
	Proceratophrys avelinoi
	Proceratophrys bagnoi
	Proceratophrys belzebul
	Proceratophrys bigibbosa
	Proceratophrys boiei
	Proceratophrys branti
	Proceratophrys brauni
	Proceratophrys caramaschii
	Proceratophrys carranca
	Proceratophrys concavitympanum

(continued)

Table 2.1 (continued)

Family	Species
	Proceratophrys cristiceps
	Proceratophrys cururu
	Proceratophrys dibernadoi
	Proceratophrys fryi
	Proceratophrys gladius
	Proceratophrys goyana
	Proceratophrys huntingtoni
	Proceratophrys itamari
	Proceratophrys izecksohni
	Proceratophrys laticeps
	Proceratophrys mantiqueira
	Proceratophrys melanopogon
	Proceratophrys minuta
	Proceratophrys moehringi
	Proceratophrys moratoi
	Proceratophrys palustris
	Proceratophrys paviotii
	Proceratophrys phyllostomus
	Proceratophrys pombali
	Proceratophrys redacta
	Proceratophrys renalis
	Proceratophrys rondonae
	Proceratophrys rotundipalpebra
	Proceratophrys sanctaritae
	Proceratophrys schirchi
	Proceratophrys strussmannae
	Proceratophrys subguttata
	Proceratophrys tupinamba
	Proceratophrys vielliardi
Phyllomedusidae	*Agalychnis aspera*
	Agalychnis buckleyi
	Agalychnis callidryas
	Agalychnis danieli
	Agalychnis granulosa
	Agalychnis hulli
	Agalychnis lemur
	Agalychnis litodryas
	Agalychnis medinae
	Agalychnis psilopygion
	Agalychnis spurrelli
	Callimedusa atelopoides
	Callimedusa baltea

(continued)

Table 2.1 (continued)

Family	Species
	Callimedusa duellmani
	Callimedusa ecuatoriana
	Callimedusa perinesos
	Callimedusa tomopterna
	Cruziohyla calcarifer
	Cruziohyla craspedopus
	Phasmahyla cochranae
	Phasmahyla exilis
	Phasmahyla guttata
	Phasmahyla jandaia
	Phasmahyla spectabilis
	Phasmahyla timbo
	Phrynomedusa appendiculata
	Phrynomedusa bokermanni
	Phrynomedusa fimbriata
	Phrynomedusa marginata
	Phrynomedusa vanzolinii
	Pithecopus ayeaye
	Pithecopus azureus
	Pithecopus centralis
	Pithecopus hypochondrialis
	Pithecopus megacephalus
	Pithecopus nordestinus
	Pithecopus oreades
	Pithecopus palliatus
	Pithecopus rohdei
Pipidae	*Pipa arrabali*
	Pipa aspera
	Pipa carvalhoi
	Pipa myersi
	Pipa parva
	Pipa pipa
	Pipa snethlageae
Ranidae	*Lithobates bwana*
	Lithobates palmipes
	Lithobates vaillanti
	Lithodytes lineatus
Rhinodermatidae	*Insuetophrynus acarpicus*
	Rhinoderma darwinii
	Rhinoderma rufum
Telmatobiidae	*Telmatobius arequipensis*
	Telmatobius atacamensis

(continued)

Table 2.1 (continued)

Family	Species
	Telmatobius atahualpai
	Telmatobius bolivianus
	Telmatobius brachydactylus
	Telmatobius brevipes
	Telmatobius brevirostris
	Telmatobius carrillae
	Telmatobius ceiorum
	Telmatobius chusmisensis
	Telmatobius cirrhacelis
	Telmatobius colanensis
	Telmatobius contrerasi
	Telmatobius culeus
	Telmatobius dankoi
	Telmatobius degener
	Telmatobius edaphonastes
	Telmatobius espadai
	Telmatobius fronteriensis
	Telmatobius gigas
	Telmatobius halli
	Telmatobius hauthali
	Telmatobius hintoni
	Telmatobius hockingi
	Telmatobius huayra
	Telmatobius hypselocephalus
	Telmatobius ignavus
	Telmatobius intermedius
	Telmatobius jelskii
	Telmatobius laevis
	Telmatobius laticeps
	Telmatobius latirostris
	Telmatobius macrostomus
	Telmatobius mantaro
	Telmatobius marmoratus
	Telmatobius mayoloi
	Telmatobius mendelsoni
	Telmatobius necopinus
	Telmatobius niger
	Telmatobius oxycephalus
	Telmatobius pefauri
	Telmatobius peruvianus
	Telmatobius philippii
	Telmatobius pinguiculus

(continued)

Table 2.1 (continued)

Family	Species
	Telmatobius pisanoi
	Telmatobius platycephalus
	Telmatobius punctatus
	Telmatobius rimac
	Telmatobius rubigo
	Telmatobius sanborni
	Telmatobius schreiteri
	Telmatobius scrocchii
	Telmatobius sibiricus
	Telmatobius simonsi
	Telmatobius stephani
	Telmatobius thompsoni
	Telmatobius timens
	Telmatobius truebae
	Telmatobius vellardi
	Telmatobius ventriflavum
	Telmatobius verrucosus
	Telmatobius vilamensis
	Telmatobius yuracare
	Telmatobius zapahuirensis

For a historical comparison, anuran species richness in South America increased 37% since the previous synthesis 18 years ago (i.e., 1644 species according to Duellman 1999). The exponential shape of the anuran discovery was expected since amphibian species richness have also increased exponentially at a global scale (see Köhler et al. 2008). Thus, many new anuran species will certainly be described in the upcoming years in South America, following trends reported for other taxa at the global scale (Joppa et al. 2011; Mora et al. 2011).

We found that at least 10 new anuran species are described per year in the period examined. However, this number was lower than average until the 1850s. Subsequently, a slight increase occurred from the 1860s onward, yet eventual decreases occurred in the 1910s and between the 1930s and the 1950s. Finally, an outstanding increase occurred from the 1960s onward, with a remarkable increase in the 1990s (Fig. 2.3).

By almost two centuries, descriptions of South American anurans were related to expeditions and subsequent work of European (e.g., André M. C. Duméril, Albert Günther, Auguste H. André Duméril, Franz Steindachner, Gabriel Bibron, George A. Boulenger, Johann Baptist von Spix, Marcos Jiménez de la Espada, Maximilian A.P. zu Wied-Neuwied, Oskar Boettger, Wilhelm Peters) and North American (Edward D. Cope) researchers. South American herpetologists started describing species in the first decades of 1900 (e.g., the Italian-Argentine José M. Gallardo and Avelino Barrio in Argentina, Alípio de Miranda-Ribeiro and Adolpho Lutz in Brazil,

Table 2.2 Number of genera
and species by anuran family
recorded in the South
America until mid-2017

Families	Genera	Species
Allophrynidae	1	3
Alsodidae	3	28
Aromobatidae	6	96
Batrachylidae	4	15
Brachycephalidae	2	64
Bufonidae	13	274
Calyptocephalellidae	2	5
Centrolenidae	13	158
Ceratophryidae	3	12
Craugastoridae	19	660
Cycloramphidae	3	36
Dendrobatidae	13	157
Eleutherodactylidae	4	13
Hemiphractidae	6	90
Hylidae	29	520
Hylodidae	3	46
Leptodactylidae	11	200
Microhylidae	12	74
Odontophrynidae	3	53
Phyllomedusidae	6	39
Pipidae	1	7
Ranidae	1	4
Rhinodermatidae	2	3
Telmatobiidae	1	64
Unassigned family	2	2
Total	**163**	**2623**

Ruan A. Rivero in Venezuela, Alberto Veloso and Ramón Formes in Chile, Pedro
M. Ruíz-C in Colombia, Jehan Vellard in Peru, Juan A. Rivero in Venezuela) (see
references in Duellman 1979; Duellman and Trueb 1994; Rossa-Feres et al. 2017).

The decreases in anuran species discovery in the 1910s and between the 1930s
and the 1950s also appear in the Fig. 5.5 of Duellman (1999). This period corre-
sponds to the two World Wars and is consistent for several taxa (animals, plants, and
protists) at different spatial scales (e.g., Reeder et al. 2007; Costello and Wilson
2011; Costello et al. 2012), including amphibians (Glaw and Köhler 1998; Köhler
et al. 2008). Thus, if war tends to accelerate development of sciences and technol-
ogy (e.g., nuclear power, telecommunication), it seems that wartime is detrimental
to the growth of species discovery, presumably through constraints on resources and
on the freedom for field trips (Willians 1998) that are essential to research
activities.

The increase in anuran species discovery in the 1960s coincides with the spread
of cladistic approaches mainly through the publication of the English edition of
Willi Henning's Phylogenetic Systematics in mid-1960s (Queiroz 1997), as well as

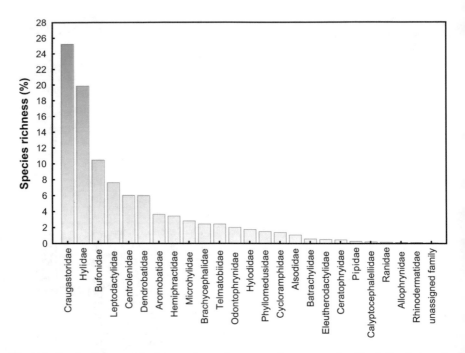

Fig. 2.1 Species richness (absolute number and percentage) per anuran families in South America until mid-2017

with the post-World War II period. The high rate of anuran species discovery from the 1990s onward can be related to multiple events, such as the spread of modern taxonomy, higher number of resident herpetologists in South America, the availability of new sampling techniques, modern travel to access remote areas, data sharing, and molecular tools.

Historical trends in the number of researchers authoring a given species description in South American increased over time (Fig. 2.4).

The increase in collaboration among researchers more recently brings multiple advantages, because it (i) enables the development of hybrid knowledge production systems, through integration of different knowledge areas, such as field and laboratory research (Vermeulen et al. 2013), (ii) gives opportunity to address complex projects that attract funding agencies and share costs in laboratory studies (see references in Nabout et al. 2015; Guerra et al. 2018), and (iii) can result in high-impact publications, including higher number of citations (Hsu and Huang 2011). In herpetology, recent uses of molecular tools increased the collaboration between traditional taxonomists and geneticists, enabling the recognition of morphologically cryptic lineages as different species (e.g., Haga et al. 2017).

Two major hotspots of newly described species are depicted when historical records of anuran species description are mapped (Fig. 2.5): (i) the Central and Northern Andes mountains and their adjacent western Amazon basin (notedly in Ecuador, Peru, and western Brazil) and (ii) the complex region encompassing the Atlantic Brazilian coast and the central Brazilian shield.

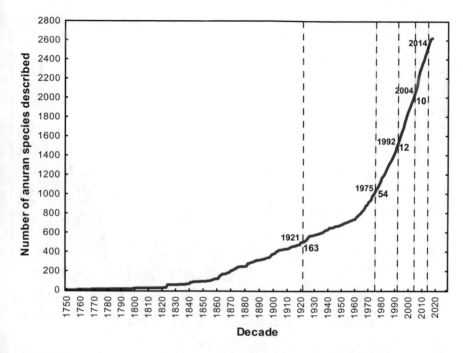

Fig. 2.2 Cumulative number of anuran species described from 1758 to mid-2017 in South America. Vertical dotted lines indicate years when multiples of 500 species were reached, with respective years. Numbers below the cumulative line (right) indicate the time span between milestones

The Amazon domain shelters one of the highest anuran diversities in South America, as it will be shown in the next chapters. Its western part present high rainfall and low seasonality, as well as high anuran diversity and endemism rates (Duellman 1999), in which a number of complexes cryptic species have been recently recorded (e.g., Ferrão et al. 2016; Caminer et al. 2017; Rojas et al. 2018). A similar pattern is recorded in the Andes as well, where many new species have been discovered (mainly craugastorids and hylids) in the wet, high-altitude forests (e.g., River-Correa et al. 2015; De la Riva and Aparicio 2016; Navarrete et al. 2016; Rodriguez and Catenazzi 2017). The second region having recently described species in South America encompasses much of the mountains of the Atlantic Forest domain, in which many anuran species recently described have restricted ranges associated with mountains (e.g., Ribeiro et al. 2015; Roberto et al. 2017). A common feature shared by the two major regions identified above is the concentration of recently described species in rough topographies, such as the Andes mountain complex and, in a lesser extent, the Atlantic Forest mountains. It is widely known that mountains act as barriers to dispersal, thus favoring allopatric speciation (Janzen 1967; Ruggiero and Hawkins 2008; Antonelli et al. 2009). Besides, mountain areas can promote adaptive speciation by parapatric and sympatric processes (Vences and Wake 2007) due to differential selection along environmental gradients (e.g.,

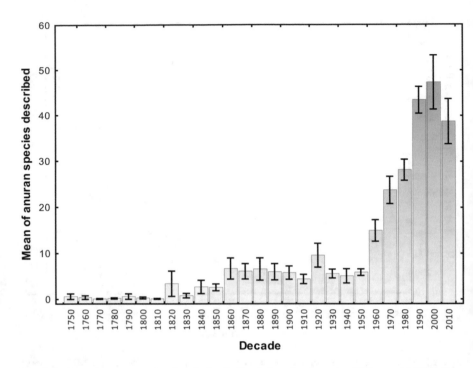

Fig. 2.3 Rate of anuran species description in South America from the 1750s to mid-2017. Bars and whiskers represent mean and standard errors by decade, respectively

Rahbek 1997; Moritz et al. 2000; Jetz et al. 2004). Indeed, the Andes mountains have been recently identified as showing great evolutionary importance and as an important source for the whole South America biodiversity (Rangel et al. 2018). Haddad and Prado (2005) also highlighted the importance of mountains for the evolution of anuran reproductive modes and species discoveries in the Atlantic Forest. Then, we expect that these South American mountain complexes (i.e., the Andes and mountain chains in the Atlantic Forest) have also an expressive number of anurans yet to be described.

Some coldspots of newly described species (i.e., areas having few species recently described) were also detected (Fig. 2.5): (i) the Llanos of Venezuela and Colombia; (ii) the northern Brazilian Amazonia; (iii) the eastern-dry diagonal between Amazon and Cerrado-Caatinga-Chaco regions (from Brazil to northeastern Bolivia), (iv) the southern Andean region (in Bolivia, Peru, Argentina, and Chile), including the Atacama Desert; and (v) the Patagonia and southern Pampean-Monte region in Argentina. Multiple causes may explain the low rates in species discovery in these coldspots, such as the historic of environmental restrictions to the occurrence of anurans and sampling biases due to accessibility (e.g., rivers, high human concentration, and/or lack of roads). Patagonian and the Southern Pampean-Monte region, the Atacama Desert, Llanos, and the "dry diagonal" are regions characterized by seasonal droughts (Cei 1979; Rivero-Blanco and Dixon 1979). Then, only

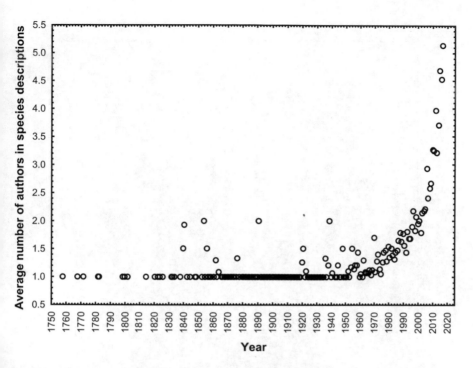

Fig. 2.4 Annual average number of authors of papers describing anuran species in South America from 1750s to mid-2017

few and highly adapted species are recorded in regions with such climatic regimes (Duellman 1999). Conversely, the possibility of sampling biases affecting the low rates in species description may be the case, at least, for the highlands of southern Andes (at the boundary between Bolivia-Peru and Argentina-Chile), as well as for the northern part of Brazilian Amazonia, where the difficult access to remote areas could generate subsampling (i.e., knowledge gaps) (Peloso 2010). These results can help direct future research effort to areas with potential occurrence of new species. An alternative would be the application of niche modeling methods of species and/ or genera endemic to these coldspots in order to guide field expeditions to those areas indicated by the models as having high climatic suitability according to the available records (e.g., Raxworthy et al. 2003).

2.3.1   Concluding Remarks

The results reported here show that the amazing anuran diversity in South America is still far from fully known. Indeed, the integration of molecular and phenotypic data has shown that the anuran diversity is largely underestimated in tropical forests of South America (Fouquet et al. 2012; Funk et al. 2012) and that some widely

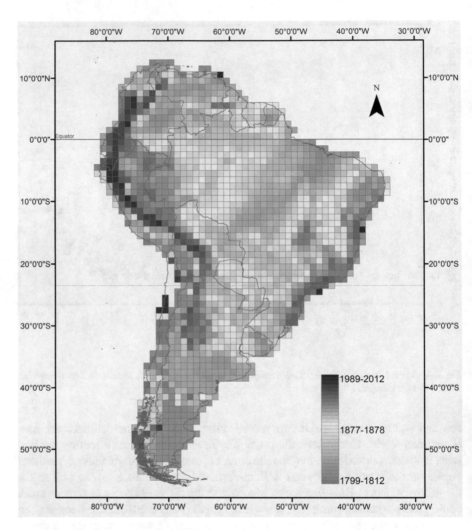

Fig. 2.5 Geographic patterns of the anuran average year of description in South America considering the geometrical interval classification from the 1750s to mid-2017. Hot colors represent more recent anuran descriptions

ranged species actually represent species complexes (see examples in Simões et al. 2010; Gehara et al. 2013, 2014, 2017). Collaborative work in species descriptions is fairly common now along with the use of integrative approaches (e.g., genetic, morphological, and bioacoustic data, as well as ecological niche models and natural history data) that help in revealing cryptic species in highly diverse regions of South America (e.g., Jansen et al. 2011; Brusquetti et al. 2014; Caminer and Ron 2014; Ferrão et al. 2016; Ortega-Andrade et al. 2015; Fouquet et al. 2016, 2018; Vacher et al. 2017). A similar scenario highlighting the power of integrative taxonomic assessments (i.e., using combined datasets of bioacoustics, morphology, and genetic) to detect cryptic diversity was reported for the Malagasy frogs (Vieites et al. 2009). In this study, the authors concluded that the number of undescribed anuran species worldwide is underestimated and that especially in tropical regions it could reach at an unprecedented level (Vieites et al. 2009). Therefore, the major centers of diversification and endemism of anurans in South America (as you will see in the next chapters) – the Amazon basin, Andean slopes, and the Atlantic Forest – will certainly remain as key regions for herpetologists to explore and uncover their amazing anuran diversity for the upcoming future. Knowing our biodiversity and the mechanisms that maintain its dynamic functioning is key to, among others, the life quality of human population. We hope that biodiversity documentations come faster than the multitude of biodiversity threats that our planet has been experiencing today (e.g., habitat loss and fragmentation, overexploitation, climate change), so we emphasize the need for higher political incentives that value basic natural history research.

Acknowledgments The authors have been continuously supported by research grants and/or fellowships from the Conselho Nacional de Desenvolvimento Científico e Tecnológico (CNPq 2037/2014-9; 431012/2016-4; 308687/2016-17; 114613/2018-4), Fundação de Amparo à Pesquisa do Estado de São Paulo (FAPESP 2011/18510-0; 2013/50714-0; 2016/13949-7), and University Research and Scientific Production Support Program of the Goias State University (PROBIP/UEG). Prof. Dr. Igor Luis Kaefer (UFAM) read critically the first version of this manuscript and provided insightful comments that improved it. We are also grateful to Brena Gonçalves Silva (UNIPAMPA) and Guilherme Castro Franco (UNIPAMPA) for their help in compiling data.

References

Antonelli A, Nylander JAA, Persson C, Sanmartin I (2009) Tracing the impact of the Andean uplift on Neotropical plant evolution. PNAS 106:9749–9754. https://doi.org/10.1073/pnas.0811421106

Antonelli A, Ariza M, Albert J, Andermann T, Azevedo J, Bacon C, Faurby S, Guedes T, Hoorn C, Lohmann LG, Matos-Maraví P, Ritter CD, Sanmartín I, Silvestro D, Tejedor M, ter Steege H, Tuomisto H, Werneck FP, Zizka A, Edwards SV (2018) Conceptual and empirical advances in Neotropical biodiversity research. PeerJ 6:e5644. https://doi.org/10.7717/peerj.5644

Brusquetti F, Jansen M, Barrio-Amarós C, Segalla M, Haddad CFB (2014) Taxonomic review of *Scinax fuscomarginatus* (Lutz, 1925) and related species (Anura; Hylidae). Zool J Linn Soc 171:783–821

Caminer MA, Ron SR (2014) Systematics of treefrogs of the *Hypsiboas calcaratus* and *Hypsiboas fasciatus* species complex (Anura, Hylidae) with the description of four new species. Zookeys 370:1–68

Caminer MA, Milá B, Jansen M, Fouquet A, Venegas PJ, Chávez G, Lougheed SC, Ron SR (2017) Systematics of the *Dendropsophus leucophyllatus* species complex (Anura: Hylidae): Cryptic diversity and the description of two new species. PLoS One 12(4):e0176902

Cei JM (1979) The Patagonian herpetofauna. In: Duellman WE (ed) The South American herpetofauna: its origin, evolution, and dispersal, vol 7. Museum of Natural History, The University of Kansas, Monography, Lawrence, pp 309–339

Costello MJ, May RM, Stork NE (2013) Can we name Earth's species before they go extinct? Science 339(6118):413–416

Costello MJ, Wilson S (2011) Predicting the number of known and unknown species in European seas using rates of description. Glob Ecol Biogeogr 20:319–330

Costello MJ, Wilson S, Houlding B (2012) Predicting total global species richness using rates of species description and estimates of taxonomic effort. Syst Biol 61(5):871–883

De la Riva I, Aparicio J (2016) Three new Bolivian species of *Psychrophrynella* (Anura: Craugastoridae), and comments on the amphibian fauna of the Cordillera de Apolobamba. Salamandra 52(4):283–292

Duellman WE (1979) The South American herpetofauna: a panoramic view. In: Duellman WE (ed) The South American herpetofauna: its origin, evolution, and dispersal, vol 7. Museum of Natural History, The University of Kansas, Monograph, Lawrence, pp 1–28

Duellman WE (1999) Distribution patterns of amphibians in South America. In: Duellman WE (ed) Patterns of distribution of amphibians: a global perspective. The Johns Hopkins University Press, Baltimore, pp 255–238

Duellman WE, Trueb L (1994) Biology of amphibians. The Johns Hopkins University Press, Baltimore

Ferrão M, Colatreli O, de Fraga R, Kaefer IL, Moravec J, Lima AP (2016) High species richness of *Scinax* treefrogs (Hylidae) in a threatened Amazonian landscape revealed by an integrative approach. PLoS One 11(11):1–16

Fouquet A, Loebmann D, Castroviejo-Fisher S, Padial JM, Orrico VGD, Lyra ML, Roberto IJ, Kok PJR, Haddad CFB, Rodrigues MT (2012) From Amazonia to the Atlantic forest: molecular phylogeny of Phyzelaphryninae frogs reveals unexpected diversity and a striking biogeographic pattern emphasizing conservation challenges. Mol Phylogenet Evol 65(2):547–561

Fouquet A, Martinez Q, Courtois EA, Dewynter M, Pineau K, Gaucher P, Blanc M, Marty C, Kok PJR (2013) A new species of the genus *Pristimantis* (Amphibia, Craugastoridae) associated with the moderately elevated massifs of French Guiana. Zootaxa 3750(5):569–586

Fouquet A, Martinez Q, Zeidler L, Courtois EA, Gaucher P, Blanc M, Lima JD, Souza SM, Rodrigues MT, Kok PJ (2016) Cryptic diversity in the *Hypsiboas semilineatus* species group (Amphibia, Anura) with the description of a new species from the eastern Guiana Shield. Zootaxa 4084(1):79–104. https://doi.org/10.11646/zootaxa.4084.1.3

Fouquet A, Vacher JP, Courtois EA, Villette B, Reizine H, Gaucher P, Jairam R, Ouboter P, Kok PJR (2018) On the brink of extinction: two new species of *Anomaloglossus* from French Guiana and amended definitions of *Anomaloglossus degranvillei* and *A. surinamensis* (Anura: Aromobatidae). Zootaxa 4379(1):1–23. https://doi.org/10.11646/zootaxa.4379.1.1

Frost DR (2017) Amphibian species of the world: an online reference. Version 6.0 Electronic Database accessible at http://research.amnh.org/herpetology/amphibia/index.html. American Museum of Natural History, New York

Funk WC, Caminer M, Ron SR (2012) High levels of cryptic species diversity uncovered in Amazon frogs. Proc R Soc 279(1734):1806–1814

Gehara M, Canedo C, Haddad CFB, Vences M (2013) From widespread to microendemic: molecular and acoustic analyses show that *Ischnocnema guentheri* (Amphibia: Brachycephalidae) is endemic to Rio de Janeiro, Brazil. Conserv Genet 14(5):973–982

Gehara M, Crawford AJ, Orrico VGD, Rodríguez A, Lötters S, Fouquet A, Barrientos LS, Brusquetti F, De la Riva I, Ernst R, Urrutia GG, Glaw F, Guayasamin JM, Hölting M, Jansen M, Kok PJR, Kwet A, Lingnau R, Lyra M, Moravec J, Pombal JP Jr, Rojas-Runjaic FJM, Schulze A, Señaris JC, Solé M, Rodrigues MT, Twomey E, Haddad CFB, Vences M, Köhler J (2014) High levels of diversity uncovered in a widespread nominal taxon: continental phylogeography of the Neotropical tree frog *Dendropsophus minutus*. PLoS One 9(9):e103958

Gehara M, Barth A, Oliveira EF, Costa MA, Haddad CFB, Vences M (2017) Model-based analyses reveal insular population diversification and cryptic frog species in the *Ischnocnema parva* complex in the Atlantic Forest of Brazil. Mol Phylogenet Evol 112:68–78

Glaw F, Köhler J (1998) Amphibian species diversity exceeds that of mammals. Herpetol Rev 29(1):11–12

Guerra V, Llusia D, Gambale PG, Morais AR, Márquez R, Bastos RP (2018) The advertisement calls of Brazilian anurans: Historical review, current knowledge and future directions. PLoS One 13(1):e0191691

Haddad CFB, Prado CPA (2005) Reproductive modes infrogs and their unexpected diversity in the Atlantic Forest of Brazil. BioScience 55:207–217

Haga IA, Andrade FS, Bruschi DP, Recco-Oimentel SM, Giaretta AA (2017) Unrevealing the leaf frogs Cerrado diversity: a new species of *Pithecopus* (Anura, Arboranae, Phyllomedusidae) from the Mato Grosso state, Brazil. PLOS One 12:e0184631

Hsu J, Huang D (2011) Correlation between impact and collaboration. Scientometrics 86:317–324

Janzen DH (1967) Why mountain passes are higher in the tropics. Am Nat 101:233–249

Jansen M, Bloch R, Schulze A, Pfenninger M (2011) Integrative inventory of Bolivia's lowland anurans reveals hidden diversity. Zool Scr 40:567–583

Jenkins CN, Pimm SL, Joppa LN (2013) Global patterns of terrestrial vertebrate diversity and conservation. PNAS 110(28):e2602–e2610

Jetz W, Rahbek C, Colwell RC (2004) The coincidence of rarity and richness and the potential signature of history in centers of endemism. Ecol Lett 7:1180–1191

Joppa LN, Roberts DL, Pimm SL (2011) The population ecology and social behaviour of taxonomists. Trends Ecol Evol 26:551–553

Köhler J, Glaw F, Vences M (2008) Essay 1.1. Trends in rates of amphibian species descriptions. In: Stuart SN, Hoffmann M, Chanson JS, Cox NA, Berridge RJ (eds) Threatened amphibians of the world. Lynx Edicions, Barcelona, p 18

Linnaeus C (1758) Systema Naturae per Regna Tria Naturae, Secundum Classes, Ordines, Genera, Species, cum Characteribus, Differentiis, Synonymis, Locis. Stockholm, Sweden

May RM (1988) How Many Species are There on Earth? Science 241:1441–1449

Mora C, Tittensor DP, Adl S, Simpson AGB, Worm B (2011) How Many Species Are There on Earth and in the Ocean? PLoS Biol 9:e1001127

Moritz C, Patton JL, Schneider CJ, Smith TB (2000) Diversification of rainforest faunas - an integrated molecular approach. Annu Rev Ecol Syst 31(1):533–563

Nabout J, Carneiro F, Borges P, Machado K, Huszar V (2015) Brazilian scientific production on phytoplankton studies: national determinants and international comparisons. Braz J Biol 75:216–223

Navarrete MJ, Venegas PJ, Ron SR (2016) Two new species of frogs of the genus *Pristimantis* from Llanganates National Park in Ecuador with comments on the regional diversity of Ecuadorian *Pristimantis* (Anura, Craugastoridae). ZooKeys 593:139–162

Oliveira EA, Rodrigues LR, Kaefer IL, Pinto KC, Hernández-Ruz EJ (2017) A new species of *Pristimantis* from eastern Brazilian Amazonia (Anura, Craugastoridae). Zookeys 687:101–129

Ortega-Andrade HM, Rojas-Soto OR, Valencia JH, Espinosa de los Monteros A, Morrone JJ, Ron SR, Canatela DC (2015) Insights from integrative systematics reveal cryptic diversity in *Pristimantis* frogs (Anura: Craugastoridae) from the Upper Amazon Basin. Plos One 10(11):e0143392

Peloso PLV (2010) A safe place for amphibians? A cautionary tale on the taxonomy and conservation of frogs, caecilians, and salamanders in the Brazilian Amazonia. Is the Brazilian Amazonia a safe place for amphibians? Or is the occurrence of threatened species in the region greatly underestimated? Zoologia 27(5):667–673

Queiroz K (1997) The Linnaean hierarchy and the evolutionization of taxonomy, with emphasis on the problem of nomenclature. Aliso 15(2):125–144

Rahbek C (1997) The relationship among area, elevation, and regional species richness in Neotropical birds. Am Nat 149(5):875–902

Rangel TF, Edwards NR, Holden PB, Diniz-Filho JAF, Gosling WD, Coelho MTP, Cassemiro FAS, Rahbek C, Colwell RK (2018) Modeling the ecology and evolution of biodiversity: biogeographical cradles, museums, and graves. Science 361:eaar5452. https://doi.org/10.1126/science.aar5452

Raxworthy CJ, Martinez-Meyer E, Horning N, Nussbaum RA, Schneider GE, Ortega-Huerta MA, Peterson AT (2003) Predicting distributions of known and unknown reptile species in Madagascar. Nature 426:837–841

Reeder DM, Helgen KM, Wilson DE (2007) Global trends and biases in new mammal species discoveries. Occasional papers, vol 269. Museum of Texas Tech University, Lubbock, pp 1–35

Ribeiro LF, Bornschein MR, Belmonte-Lopes R, Firkowski CR, Morato SAA, Pie MR (2015) Seven new microendemic species of *Brachycephalus* (Anura: Brachycephalidae) from southern Brazil. PeerJ 3:e1011. https://doi.org/10.7717/peerj.1011

River-Correa M, Burneo KG, Grant T (2015) A new red-eyed of stream treefrog of *Hyloscirtus* (Anura: Hylidae) from Peru, with comments on the taxonomy of the genus. Zootaxa 4061(1):29–40

Rivero-Blanco C, Dixon JR (1979) Origin and distribution of the herpetofauna of the dry lowland tropical rainforests of South America. In: Duellman WE (ed) The South American Herpetofauna: its origin, evolution, and dispersal, vol 7. Museum of Natural History, The University of Kansas, Monography, Lawrence, pp 281–298

Roberto IJ, Araujo-Vieira K, Carvalho-e-Silva SP, Ávila RW (2017) A New Species of *Sphaenorhynchus* (Anura: Hylidae) from Northeastern Brazil. Herpetologica 73(2):148–161

Rodriguez LO, Catenazzi A (2017) Four new species of terrestrial-breeding frogs of the genus *Phrynopus* (Anura: Terrarana: Craugastoridae) from Río Abiseo National Park, Peru. Zootaxa 4273(3):381–406

Rojas RR, Fouquet A, Ron SR, Hernández-Ruz EJ, Melo-Sampaio PR, Chaparro JC, Vogt RC, Carvalho VT, Pinheiro LC, Avila RW, Farias IP, Gordo M, Hrbek T (2018) A Pan-Amazonian species delimitation: high species diversity within the genus *Amazophrynella* (Anura: Bufonidae). PeerJ 6:e4941. https://doi.org/10.7717/peerj.4941

Rossa-Feres D d C, Garey MV, Caramaschi U, MF NI, Nomura F, Bispo A, Brasileiro CA, MTC T, Sawaya RJ, Conte CE, Cruz CA, Nascimento LB, Gasparini JL, Almeida NP, Haddad CFB (2017) Anfíbios da Mata Atlântica: lista de espécies, histórico dos estudos, biologia e conservação. In: Monteiro-Filho EA, Conte CE (eds) Revisões em Zoologia: Mata Atlântica. Editora UFPR, Curitiba, pp 237–314

Ruggiero A, Hawkins BA (2008) Why do mountains support so many species of birds? Ecography 31:306–315. https://doi.org/10.1111/j.2008.0906-7590.05333.x

Simões PI, Lima AP, Farias IP (2010) The description of a cryptic species related to the pan-Amazonian frog *Allobates femoralis* (Boulenger 1883) (Anura: Aromobatidae). Zootaxa 2406:1–18

Vasconcelos TS, Rodríguez MA, Hawkins BA (2012) Species distribution modelling as a macroecological tool: a case study using New World amphibians. Ecography 35:539–548

Vacher JP, Martinez Q, Fallet M, Courtois E, Blanc M, Gaucher P, Dewynter M, Jairam R, Ouboter P, Thebaud C, Fouquet A (2017) Cryptic diversity in Amazonian frogs: integrative taxonomy of the genus *Anomaloglossus* (Amphibia: Anura: Aromobatidae) reveals a unique case of diversification within the Guiana Shield. Mol Phylogenet Evol 112:158–173

Vieites DR, Wollenberg KC, Andreone F, Köhler J, Glaw F, Vences M (2009) Vast underestimation of Madagascar's biodiversity evidenced by an integrative amphibian inventory. PNAS 106(20):8267–8272

Vences M, Wake DB (2007) Speciation, species boundaries and phylogeography of amphibians. In: Heatwole HH, Tyler M (eds) Amphibian Biology. Surrey Beatty & Sons, Chipping Norton, pp 2613–2669

Vermeulen N, Parker JN, Penders B (2013) Understanding life together: a brief history of collaboration in biology. Endeavour 37:162–171

Villalobos F, Dobrovolski R, Provete DB, Gouveia SF (2013) Is Rich and rare the common share? Describing biodiversity patterns to inform conservation practices for South American anurans. PLoS One 8(2):e56073. https://doi.org/10.1371/journal.pone.0056073

Willians PH (1998) An annotated checklist of bumble bees with an analysis of patterns of description (Hymenoptera: Apidae, Bombini). Bull Nat Hist Mus Entomol 67(1):79–152

Chapter 3
Patterns of Species Richness, Range Size, and Their Environmental Correlates for South American Anurans

Abstract Species richness and range size gradients have been correlated with environmental conditions at broad spatial scales, yet these effects are commonly context-dependent for different geographical regions. Here we assembled range maps of South American anurans and used spatial and nonspatial regressions to assess the potential influences of environmental variables on the gradients of species richness and range sizes. Additionally, we evaluated the consistency of these environmental drivers separately for temperate/subtropical and tropical regions of South America. We found that vegetation structure, temperature, and energy-water balance were the strongest predictors of species richness at the continental scale; temperature, productivity, and elevation were the best predictors for range size. Explanatory power of predictors shifted across different regions of the continent: in the tropical, vegetation structure was the strongest correlate of species richness, and in the temperate/subtropical, temperature and energy-water balance were the most important predictors. As for range size, elevation and temperature were the best predictors in the tropical region, whereas temperature seasonality was the strongest predictor in the temperate/subtropical region. Our results support the idea that different environmental filters can vary according to the latitude, reinforcing the relevance of evaluating patterns at multiple spatial scales to understand environmental drivers of biodiversity.

Keywords Amphibians · Climate variability · Energy water · Environmental gradients · Range size · Species diversity · Non stationarity · Autoregressive models · Hierarchical partitioning

3.1 Introduction

Species richness and geographic ranges are fundamental unities to understand patterns and processes that govern the biodiversity on Earth (Brown et al. 1996; Hawkins et al. 2003). Species richness refers to the number of species in a given

© Springer Nature Switzerland AG 2019
T. S. Vasconcelos et al., *Biogeographic Patterns of South American Anurans*,
https://doi.org/10.1007/978-3-030-26296-9_3

area. Elucidating the mechanisms governing its variation across an area is one of the central issues in macroecology and biogeography (Gotelli et al. 2009). Gradients of species richness at broad scales have long been attributed to variations in climatic/ productivity variables (Hawkins et al. 2003), yet other non-climatic variables have been recognized to be important correlates of species richness (e.g., evolutionary history: Ricklefs 2004; Field et al. 2009; Marin and Hedges 2016), including geo- metrical constraints (Colwell and Lees 2000; Colwell et al. 2016), some of them mainly recognized at finer spatial scales (Field et al. 2009).

The geographic range size of species (hereafter range size) is generally linked to the latitudinal gradients in species richness (e.g., Gaston et al. 2008). For instance, the *climate variability hypothesis* states that organisms living in cold, extratropical climates have high tolerance to harsh environmental conditions, which in turn results in larger range sizes at high latitudes than their counterparts in the Tropics (Stevens 1989). This mechanism could explain the pattern often called Rapoport's rule (Rapoport 1982), which states that range sizes correlate positively with latitude, resulting in lower species richness with larger range sizes at high latitudes and higher species richness with narrower ranges in the Tropics. Rapoport's rule has been reported for amphibians in the northern hemisphere (Whitton et al. 2012), but range size patterns may become more complex depending on taxa and geographic regions, such as South American birds and their complex correlation with habitat, topography, and climate regimes across the continent (Hawkins and Diniz-Filho 2006). Moreover, other environmental factors may influence range size, such as energy input and productivity that theoretically favor greater population densities (Yee and Juliano 2007), which in turn may lead to a reverse Rapoport pattern in the Tropics (Whitton et al. 2012).

Examining the effects of environmental factors on patterns of species richness and range sizes on a regional basis is important because context-dependent effects (non-stationarity) have been found across multiple geographical regions (Cassemiro et al. 2007; Powney et al. 2010; Buschke et al. 2015). For instance, water availabil- ity usually is the strongest predictor of species richness in the Tropics and subtrop- ics, while energy variables are dominant at higher latitudes (Hawkins et al. 2003). Another example is that Rapport's rule is applicable mainly for organisms in the northern hemisphere, but may reverse for the southern hemisphere (Whitton et al. 2012).

Documentations of amphibian richness gradients and range size patterns have been performed in North America, Europe, Australia, Africa, and China (VanDerWal et al. 2008; Escoriza and Ruhí 2014; Luo et al. 2015; Trakimas et al. 2016; Lewin et al. 2016; Coops et al. 2018). In South America, these macroecological patterns have also been explored for amphibians, either at the continental (Vasconcelos et al. 2012; Villalobos et al. 2013; Moura et al. 2016) or global scales (Buckley and Jetz 2007; Fritz and Rahbek 2012). However, none of them modeled these geographical patterns as a function of their environmental correlates in the context-dependent effect across different geographic regions.

Here, we map the species richness gradient and the range size patterns for South American anurans with an updated species list detailed in Chaps. 1 and 2. Additionally, we identify, through regression and correlation analyses, which environmental variables (i.e., climate, productivity, topography, and habitat structure variables) better correlate with these two macroecological patterns. Finally, we repeat this same approach to identify specific environmental correlates for tropical and temperate/subtropical regions, separately.

3.2 Material and Methods

3.2.1 Anuran Species Richness and Range Sizes

We used the anuran species distribution dataset described in Chap. 1. In summary, each species map was overlaid into the South America grid system with resolution of ~100 km. Then, the species richness per grid cell was taken by summing up all species occurring at each grid. The range size values were calculated based on the number of grid cells occupied by the species. Then, the mean range sizes were calculated by averaging the ranges of all species present in a given grid cell.

3.2.2 Environmental Data

We selected nine environmental variables that have been shown to affect species diversity via habitat heterogeneity, environmental energy, and environmental stress (Buckley and Jetz 2007; Whitton et al. 2012; Moura et al. 2016). The climatic variables selected were mean annual temperature (TEMP), temperature seasonality (TSEASO), mean annual precipitation (PRECIP), and precipitation seasonality (PRECSEASO). Elevation range (ELEV) was considered as a measure of topographic heterogeneity, annual actual evapotranspiration (AET) as a measure of water-energy balance, and the Normalized Difference Vegetation Index (NDVI) as a measure of primary productivity. Mean canopy height (CANOP) and standard

Table 3.1 Biomes in South America (Olson et al. 2001) and their classification as tropical or temperate according to their latitudinal extent

Biome	Classification
Chaco	Tropical
Dry forest	Tropical
Mediterranean	Temperate/subtropical
Moist/semideciduous forest	Tropical
Montane grasslands	Tropical
Temperate forest	Temperate/subtropical
Temperate grasslands	Temperate/subtropical
Tropical savannas	Tropical
Xeric	Tropical

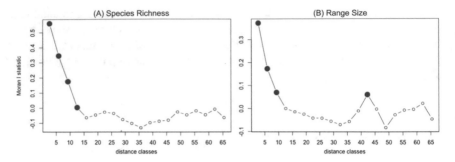

Fig. 3.1 Moran's *I* correlograms showing residual spatial autocorrelation along distance classes (in decimal degrees) for the OLS full models of (**a**) species richness and (**b**) range sizes across South America. Red dots indicate significant spatial autocorrelation

deviation of canopy height (CANSD) were considered as two measures of vegetation complexity. Sources, resolution, and data handling are detailed in Chap. 1.

We also obtained for each grid cell its major biome according to the World Wildlife Fund designations (Olson et al. 2001). The biomes were then classified as temperate/subtropical or tropical according to its latitudinal extent (Table 3.1). For subsequent analysis, we evaluated the results for the total extent of South America and separately for temperate/subtropical and tropical biomes in the continent.

3.2.3 Data Analysis

Prior to performing all analyses, species richness, mean range size, annual mean precipitation, elevational range, and mean canopy height were square root transformed, whereas temperature seasonality and precipitation seasonality were log transformed to reduce heteroscedasticity and normalize model residuals. We standardized each predictor variable using z-scores (Quinn and Keough 2002) to provide comparable regression coefficients. All environmental variables were tested for collinearity using the variance inflation factors (VIF), and only those with low collinearity were retained in the model (VIF $\leq$ 5.03; Quinn and Keough 2002).

We used Pearson's correlation to measure the relationship between species richness and average range sizes, and between these variables and latitude, with Clifford's correction of degrees of freedom (Clifford et al. 1989) to account for spatial correlation. We assessed the independent contributions of each environmental variable on species richness and range sizes using hierarchical partitioning (Mac Nally 2002).

To assess models that best explain the relationship between species richness, range size, and environmental predictors, we fitted general linear models (ordinary least squares, OLS). The presence of spatial autocorrelation in OLS model residuals violates the assumption of residual independence and can distort estimates of model parameters. Then, we generated spatial correlograms using Moran's I to assess autocorrelation in species richness and range sizes (OLS residuals) as a function of the distance classes. Because a spatial structure was evident in OLS residuals (Fig. 3.1), we fitted spatially explicitly simultaneous autoregressive models (SAR, Kissling and Carl 2008) – which allow the inclusion of the residual spatial autocorrelation of the data. We defined as neighborhood points separated by ~200 km in SAR analyses because this distance corresponded to the first class in the spatial correlograms, which displayed the strongest spatial autocorrelation for both species richness and range size (Fig. 3.1).

OLS and SAR models for species richness and range sizes were carried out for all possible combinations of environmental predictors including a model containing only the intercept. Model selection procedure was implemented using the Akaike Information Criteria corrected for small samples (AICc) to compare the fit of competing models (Burnham and Anderson 2002). A model was regarded as unsupported if its AICc weight was less than 1/8 of the best model (Burnham and Anderson 2002). We also used AIC weights to estimate parameter values and their variances, which influence both species richness and range size. This analysis was based on a model-averaging approach (Burnham and Anderson 2002). The model-averaged estimates and standard errors (SE) were calculated, and 95% confidence intervals were employed to assess the magnitude of the effect. We concluded that there was an effect if the confidence interval excluded 0 (Werner et al. 2007). We also calculated pseudo-R^2 values from our OLS and SAR averaged models as a squared Pearson correlation between predicted and observed values (Kissling and Carl 2008). All statistical analyses were performed in R v. 3.4.1 (R Development Core Team 2017), using the packages "hier.part," "SpatialPack," "spdep," and "MuMIn."

3.3 Results

3.3.1 Patterns of Species Richness and Range Sizes

Species richness and range sizes were marginally correlated ($r = 0.51$, $P = 0.06$) across South America. Species richness varied from 1 to 184 species per grid cell ($\bar{x} = 50.07 \pm 34.94$) and is clearly higher in tropical South America than at higher latitudes (Fig. 3.2a), though a robust latitudinal gradient is not evident ($r = 0.64$, $P = 0.095$) probably due to the complex richness distribution within the tropical region: higher anuran diversity is found in tropical forests irrespective of their latitudinal locations (e.g., higher diversity in the Amazon and Atlantic forests), and lower diversity is found in tropical open formations (e.g., the Cerrado-Caatinga-Chaco diagonal in between the Amazon and Atlantic forests; Fig. 3.2a). Conversely, species richness reaches their lowest values in the temperate Patagonian and Andes mountain complex (Fig. 3.2a).

Mean range sizes varied from 10,000 km² to 7,967,500 km² per grid cell ($\bar{x} = 3,369,700$ km² $\pm 1,685,309$) and does not exhibit a latitudinal gradient ($r = 0.50$, P

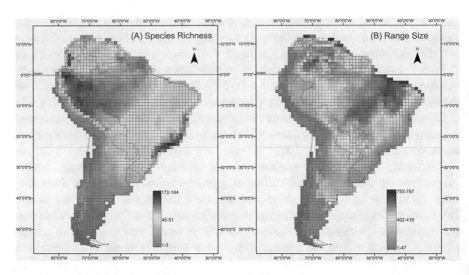

Fig. 3.2 Spatial distribution of anuran (**a**) species richness and (**b**) mean range size in $1.0^\circ \times 1.0^\circ$ grid cells across South America

= 0.173): small-ranged species occur mostly in the Patagonia but also along the Andes mountain complex and the coastal Atlantic forest (Fig. 3.2b). Conversely, wide-ranged species occur mostly in lowlands of the Amazonia and Cerrado-Caatinga-Chaco diagonal (Fig. 3.2b).

3.3.2   Environmental Correlates of Species Richness and Range Sizes

SAR models had lower AICc values from the best model, higher number of models in the confidence set, and higher pseudo-R^2 values (Tables 3.2 and 3.3). Model average coefficients for both OLS and SAR models showed little divergence in the direction of the effects, but OLS coefficients were usually higher than SAR coefficients (Tables 3.2 and 3.3).

CANOP and AET were consistently recognized as the strongest predictors of species richness in both OLS and SAR models and in hierarchical partitioning (Table 3.2, Fig. 3.3a). Additionally, TEMP was also a strong predictor of species richness in the OLS models and hierarchical partitioning (Table 3.2, Fig. 3.3a), and TSEASO was the second most important predictor in SAR models (Table 3.2). TEMP, NDVI, and ELEV were among the strongest predictors of range size in OLS models and hierarchical partitioning (Table 3.3, Fig. 3.3b). Alternatively, the SAR models showed AET and TSEASO as the best predictors of range sizes (Table 3.3).

The environmental variables influencing species richness and range size varied between the temperate/subtropical and tropical regions of South America. In the

Table 3.2 Results of nonspatial ordinary least squares (OLS) and spatially explicit simultaneous autoregressive (SAR) models to explain variation in anuran species richness across all South America and its temperate/subtropical and tropical regions

	All South America		Temperate/subtropical South America		Tropical South America	
	OLS models	SAR models	OLS models	SAR models	OLS models	SAR models
Number of models in the confidence set	9	3	12	38	4	2
AICc of the best model	5430.1	2922.2	387.8	209.3	4530.4	2556.6
Support for the best model (*wi*)	0.20	0.54	0.17	0.06	0.46	0.56
Averaged model coefficient ± SE						
Intercept	**6.47 ± 0.03**	**6.34 ± 0.65**	**2.25 ± 0.03**	**2.72 ± 0.88**	**7.30 ± 0.03**	**7.03 ± 0.74**
TEMP	**0.85 ± 0.07**	**0.23 ± 0.06**	**0.41 ± 0.06**	0.02 ± 0.06	**0.24 ± 0.06**	**0.16 ± 0.05**
TSEASO	-0.04 ± 0.06	**-0.50 ± 0.12**	0.06 ± 0.06	0.01 ± 0.05	0.02 ± 0.05	**-0.36 ± 0.12**
PRECIP	**0.16 ± 0.07**	-0.11 ± 0.07	0.10 ± 0.08	0.09 ± 0.09	**0.31 ± 0.07**	**-0.21 ± 0.06**
PRECSEASO	0.01 ± 0.03	-0.01 ±0.04	0.02 ± 0.04	**-0.22 ± 0.09**	-0.06 ± 0.05	-0.03 ± 0.05
ELEV	**0.17 ± 0.05**	**0.37 ± 0.04**	0.10 ± 0.05	0.10 ± 0.06	**-0.18 ± 0.05**	**0.39 ± 0.04**
AET	**0.81 ± 0.07**	**0.61 ± 0.05**	**0.42 ± 0.05**	0.05 ± 0.05	**0.44 ± 0.07**	**0.63 ± 0.04**
NDVI	**0.26 ± 0.06**	**0.19 ± 0.05**	**0.24 ± 0.06**	0.01 ± 0.03	**0.16 ± 0.05**	**0.20 ± 0.05**
CANOP	**0.94 ± 0.06**	**0.48 ± 0.06**	-0.01 ± 0.04	-0.11 ± 0.07	**0.97 ± 0.06**	**0.53 ±0.07**
CANOPSD	0.04 ± 0.04	**0.10 ± 0.02**	**0.23 ± 0.07**	**0.34 ± 0.08**	**0.18 ± 0.04**	**0.11 ± 0.02**
Averaged model pseudo-R^2	0.81	0.97	0.79	0.92	0.71	0.94

Note: Bolded values indicate coefficient estimates for which confidence interval excluded zero

temperate/subtropical, AET and TEMP appeared consistently among the best predictors of species richness (Table 3.2, Fig. 3.3c). We found support for the effects of TSEASO and PRECSEASO on range sizes in both OLS and SAR models (Table 3.3), and TSEASO also had the highest independent contribution to explain variation in range sizes in temperate/subtropical South America (Fig. 3.3d). In the tropical region, CANOP and AET were among the best predictors of species richness in both OLS and SAR models and in hierarchical partitioning (Table 3.2, Fig. 3.3e). TEMP, NDVI, and ELEV were among the best predictors for range sizes in both OLS and hierarchical partitioning (Table 3.3, Fig. 3.3f), whereas TEMP and

Table 3.3 Results of nonspatial ordinary least squares (OLS) and spatially explicit simultaneous autoregressive (SAR) models to explain variation in anuran range sizes across all South America and its temperate/subtropical and tropical regions

	All South America		Temperate/subtropical South America		Tropical South America	
	OLS models	SAR models	OLS models	SAR models	OLS models	SAR models
Number of models in the confidence set	1	12	9	18	2	13
AICc best model	8200.5	6163.5	1185.0	1039.4	6634.6	5115.6
Support for the best model (*wi*)	1.0	0.22	0.29	0.13	0.71	0.13
Averaged model coefficient ± SE						
Intercept	**17.5 ± 0.07**	**16.73 ± 1.75**	**11.85 ± 0.13**	**11.55 ± 1.45**	**18.61 ± 0.07**	**18.07 ± 1.21**
TEMP	**2.24 ± 0.16**	**1.09 ± 0.15**	**1.17 ± 0.26**	-0.16 ± 0.30	**1.56 ± 0.13**	**1.01 ± 0.12**
TSEASO	**-0.80 ± 0.15**	**1.29 ± 0.32**	**2.95 ± 0.20**	**1.08 ± 0.52**	**-0.66 ± 0.13**	**0.82 ± 0.29**
PRECIP	**-0.67 ± 0.15**	-0.03 ± 0.09	-0.07 ± 0.18	**-0.82 ± 0.45**	0.02 ± 0.09	0.03 ±0.09
PRECSEASO	**0.59 ± 0.10**	-0.09 ± 0.14	**-1.28 ± 0.20**	**-1.17 ± 0.37**	**1.27 ± 0.10**	0.13 ± 0.17
ELEV	**-1.56 ± 0.10**	**-0.53 ± 0.10**	**-0.69 ± 0.20**	**-0.67 ± 0.23**	**-1.40 ± 0.12**	**-0.66 ± 0.10**
AET	**0.66 ± 0.15**	**1.34 ± 0.12**	**0.76 ± 0.21**	**0.94 ± 0.23**	**0.80 ± 0.13**	**1.11 ± 0.11**
NDVI	**1.83 ± 0.14**	0.01 ± 0.08	0.37 ± 0.33	-0.01 ± 0.10	**2.0 ± 0.13**	0.11 ± 0.12
CANOP	**-1.05 ± 0.15**	0.21 ± 0.32	-0.65 ± 0.34	-0.001 ± 0.10	**-1.39 ± 0.15**	0.13 ± 0.16
CANOPSD	**-0.58 ± 0.09**	**-0.40 ± 0.06**	-0.10 ± 0.30	-0.07 ± 0.17	**-0.60 ± 0.09**	**-0.39 ± 0.06**
Averaged model pseudo-R^2	0.73	0.94	0.83	0.92	0.71	0.92

Note: Bolded values indicate coefficient estimates for which confidence interval excluded zero

AET were the best predictors in SAR models of range sizes in tropical South America (Table 3.3).

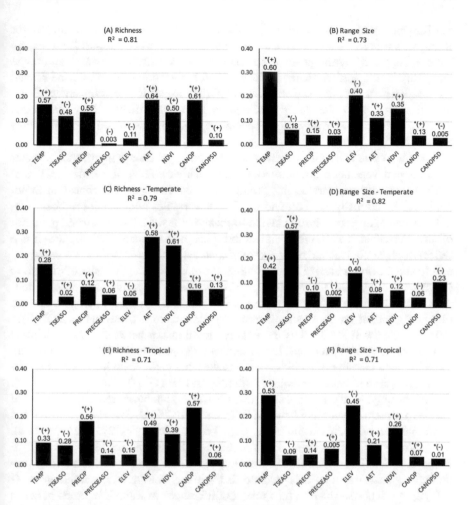

Fig. 3.3 Results of hierarchical partitioning showing the relative importance of predictors of (**a**) species richness and (**b**) range size in all South America, (**c**) species richness and (**d**) range size in temperate South America, (**e**) species richness and (**f**) range size in all South America. Numbers above bars are regression coefficients of the variable in a simple regression, (+) positive or (−) negative coefficients, *significant P-values for the variable in a simple regression

3.4 Discussion

Latitudinal gradients on species richness have been widely reported for several vertebrate groups (Rahbek et al. 2007; Jenkins et al. 2013; Lewin et al. 2016), including amphibians (Buckley and Jetz 2007; Moura et al. 2016). Mechanistic models to explain these patterns have been proposed (Whitton et al. 2012; Belmaker and Jetz 2012), such as adaptations of organisms to tolerate environmental stress, presumably affecting the relationship between species richness and range size proposed by the Rapoport's rule (Rapoport 1982). As expected, we did not find support for Rapoport's rule for South American anurans. We found more support for a reverse pattern, suggested by the positive correlation between species richness and range size. That is to say, sites richer in species had higher proportions of wide-ranged species, whereas the poorest grid cells had higher proportions of small-ranged species. However, richer sites in tropical highlands, such as Atlantic Rain Forests and Tropical Andes, had narrow-ranged species, similar to New World birds (Hawkins and Diniz-Filho 2006).

Latitude per se does not explain anuran richness, yet energy input, energy-water balance, and vegetation structure (i.e., the positive effects of TEMP, AET, and CANOP, respectively) are strong predictors of anuran species richness in South America. Additionally, the correlation of these variables with species richness varied between temperate/subtropical and tropical regions. AET was a strong predictor of anuran richness in both temperate/subtropical and tropical portions, which is a reflect of physiological and reproductive dependence of anurans on the water availability and the importance of the water-energy balance for the diversity of this group (Buckley and Jetz 2007). Conversely, TEMP was a strong predictor of species richness specially in the temperate/subtropical region. As ectotherms, amphibians are highly dependent on the environmental temperature to activate their metabolism (Vitt and Caldwell 2009); thus the positive relationship between anuran richness and temperature was expected. In temperate/subtropical regions, temperatures are more extreme than in the Tropics, which makes this variable an important filter for assembling amphibian communities (Buckley and Jetz 2007).

Surprisingly, we did not find significant correlation between rainfall regimes and species richness (represented by PRECIP or PRECSEASO), which is a pattern commonly reported in the literature (e.g., Vasconcelos et al. 2010; Moura et al. 2016). Water per se is represented in this study by AET and indirectly by CANOP. We found that CANOP, a variable representing vegetation complexity, is a strong predictor of species richness. This predictor is positively correlated to rainfall values ($r = 0.72$, $P = 0.004$), which in turn reflects differences in rainfall regimes between moist forests and open areas/dry forests. Moist forests provide microhabitat availability for foraging, shelter, and diversification of reproductive modes for anurans (Haddad and Prado 2005; da Silva et al. 2012; Müller et al. 2013; Vági et al. 2019), being thus considered one of the most important drivers of anuran diversification in the continent (Haddad and Prado 2005; Crump 2015).

Range size was positively correlated with energy input (TEMP), and primary productivity (NDVI), and negatively with topography. Theory predicts that primary productivity increases carrying capacity of populations, allowing higher densities of organisms (Yee and Juliano 2007). Then, primary productivity can ultimately reduce extinction risks and increase range sizes (Whitton et al. 2012). Rough topographies create physical and climatic barriers that difficult dispersal among mountain chains due to niche conservatism (Wiens and Graham 2005), favoring allopatric speciation of new small-ranged species in the evolutionary context. This is expected to be more pronounced in tropical altitudes than in temperate regions, due to the accentuated climatic differences between high and lowlands (e.g., the tropical Andes and the southeastern Atlantic Forest coast) (Ghalambor et al. 2006; Polato et al. 2018), but we also found this pattern in the temperate Andes.

The environmental variables correlated with range size in the tropical region were the same detected for the whole South America. This scenario is different in the temperate/subtropical region of the continent, in which range size was correlated with temperature seasonality. These results confirm our expectations that different variables influence range size at different latitudes. For instance, at a global scale (Whitton et al. 2012), temperature seasonality was the most important predictor of amphibian range size in the Nearctic and Palearctic realms. Here, we found that temperature seasonality is important within the temperate/subtropical region of South America, following the climate variability hypothesis (CVH; Stevens 1989). This hypothesis states that species inhabiting places with more variable climate evolves broader physiological tolerances and, consequently, broader niches and wider ranges than species living in more stable regions (Stevens 1989). Although we did not find support for this hypothesis for the whole South America, CVH appears to be important at least in temperate/subtropical South America, where temperature becomes more limiting than in the tropical portion of the continent.

To our knowledge, this is the first study to test non-stationarity in species richness and range size patterns for South American anurans. Studies conducted at global scales have shown different effects of environmental variables on species richness or range size depending on the spatial scale (Buckley and Jetz 2007; Whitton et al. 2012; Gouveia et al. 2013). We found that considering different climatic regimes within the continent is important to identify the specific environmental determinants of richness gradients and geographic range patterns of anurans. Even though its climatic differences are less accentuated than in the Northern Hemisphere (Whitton et al. 2012; Sunday et al. 2014), anuran richness and range sizes appear to have different environmental determinants in temperate/subtropical and tropical South America.

Acknowledgments The authors have been continuously supported by research grants and/or fellowships from the Fundação de Amparo à Pesquisa do Estado de São Paulo (FAPESP 2011/18510-0; 2013/50714-0; 2016/13949-7), Conselho Nacional de Desenvolvimento Científico e Tecnológico (CNPq 2037/2014-9; 431012/2016-4; 308687/2016-17; 114613/2018-4), and University Research and Scientific Production Support Program of the Goias State University (PROBIP/UEG). Prof. Dr. Fabrício Barreto Teresa (UEG) read critically the first version of this manuscript and provided insightful comments that improved it.

References

Belmaker J, Jetz W (2012) Regional pools and environmental controls of vertebrate richness. Am Nat 179:512–523

Brown JH, Stevens GC, Kaufman DF (1996) The geographic range: size, shape, boundaries, and internal structure. Ann Rev Ecol Syst 27:597–623

Buckley LB, Jetz W (2007) Environmental and historical constraints on global patterns of amphibian richness. P Roy Soc B-Biol Sci 274:1167–1173

Burnham KP, Anderson DR (2002) Model selection and multimodel inference: a practical information-theoretic approach, 2nd edn. Springer, New York

Buschke FT, De Meester L, Brendonck L, Vanschoenwinkel B (2015) Partitioning the variation in African vertebrate distributions into environmental and spatial components – exploring the link between ecology and biogeography. Ecography 38:450–461

Cassemiro FAS, Barreto BS, Rangel TFLV, Diniz-Filho JAF (2007) Non-stationarity, diversity gradients and the metabolic theory of ecology. Global Ecol Biogeogr 16:820–822

Clifford P, Richardson S, Hemon D (1989) Assessing the significance of the correlation between two spatial processes. Biometrics 45:123–134

Colwell RK, Lees DC (2000) The mid domain effect: geometry constraints on the geography of species richness. Trends Ecol Evol 15:70–76

Colwell RK, Gotelli NJ, Ashton LA et al (2016) Midpoint attractors and species richness: modelling the interaction between environmental drivers and geometric constraints. Ecol Lett 19:1009–1022

Coops NC, Rickbeil GJM, el al BDK (2018) Disentangling vegetation and climate as drivers of Australian vertebrate richness. Ecography 41:1147–1160

Crump ML (2015) Anuran reproductive modes: evolving perspectives. J Herpetol 49:1–16

da Silva FR, Almeida-Neto M, Prado VHM et al (2012) Humidity levels drive reproductive modes and phylogenetic diversity of amphibians in the Brazilian Atlantic Forest. J Biogeogr 39:1720–1732

Escoriza D, Ruhí A (2014) Macroecological patterns of amphibian assemblages in the Western Palearctic: Implications for conservation. Biol Conserv 176:252–261

Field R, Hawkins BA, Cornell HV et al (2009) Spatial species-richness gradients across scales: a meta-analysis. J Biogeogr 36:132–147

Fritz SA, Rahbek C (2012) Global patterns of amphibian phylogenetic diversity. J Biogeogr 39:1373–1382

Gaston KJ, Chown SL, Evans KL (2008) Ecogeographical rules: elements of a synthesis. J Biogeogr 35:483–500

Ghalambor CK, Huey RB, Martin PR et al (2006) Are mountain passes higher in the tropics? Janzen's hypothesis revisited. Integr Comp Biol 46:5–17

Gotelli NJ, Anderson MJ, Arita HT et al (2009) Patterns and causes of species richness: a general simulation model for macroecology. Ecol Lett 12:873–886

Gouveia SF, Hortal J, Cassemiro FAZ et al (2013) Nonstationary effects of productivity, seasonality, and historical climate changes on global amphibian diversity. Ecography 36:104–113

Haddad CFB, Prado CPA (2005) Reproductive modes in frogs and their unexpected diversity in the Atlantic Forest of Brazil. BioScience 55:207–217

Hawkins BA, Diniz-Filho JAF (2006) Beyond Rapoport's rule: evaluating range size patterns of New World birds in a two-dimensional framework. Global Ecol Biogeogr 15:461–469

Hawkins BA, Field R, Cornell HV et al (2003) Energy, water, and broad-scale geographic patterns of species richness. Ecology 84:3105–3117

Jenkins CN, Pimm SL, Joppa LN (2013) Global patterns of terrestrial vertebrate diversity and conservation. P Natl Acad Sci USA 110:E2602–E2610

Kissling WD, Carl G (2008) Spatial autocorrelation and the selection of simultaneous autoregressive models. Global Ecol Biogeogr 17:59–71

Lewin A, Feldman A, Bauer AM et al (2016) Patterns of species richness, endemism and environmental gradients of African reptiles. J Biogeogr 43:2380–2390

Luo Z, Wei S, Zhang W et al (2015) Amphibian biodiversity congruence and conservation priorities in China: Integrating species richness, endemism, and threat patterns. Biol Conserv 191:650–658

Mac Nally R (2002) Multiple regression and inference in ecology and conservation biology: further comments on identifying important predictor variables. Biodivers Conserv 11:1397–1401

Marin J, Hedges SB (2016) Time best explains global variation in species richness of amphibians, birds and mammals. J Biogeogr 43:1069–1079

Moura MR, Villalobos F, Costa GC, Garcia PCA (2016) Disentangling the role of climate, topography and vegetation in species richness gradients. PLoS ONE 11:e0152468

Müller H, Liedtke HC, Menegon M et al (2013) Forests as promoters of terrestrial life-history strategies in East African amphibians. Biology Lett 9:20121146

Olson DM, Dinerstein E, Wikramanayake ED et al (2001) Terrestrial ecoregions of the world: a new map of life on Earth. BioScience 51:933–938

Polato NR, Gill BA, Shah AA et al (2018) Narrow thermal tolerance and low dispersal drive higher speciation in tropical mountains. P Natl Acad Sci USA 115:12471–12476

Powney GD, Grenyer R, Orme CDL et al (2010) Hot, dry and different: Australian lizard richness is unlike that of mammals, amphibians and birds. Global Ecol Biogeogr 19:386–396

Quinn GP, Keough MJ (2002) Experimental design and data analysis for biologists. Cambridge University Press, Cambridge, UK

R Development Core Team (2017) R: a language and environment for statistical computing. R Foundation for Statistical Computing, Vienna. http://www.R-project.org. Accessed 30 Jan 2018

Rahbek C, Gotelli NJ, Colwell RK et al (2007) Predicting continental-scale patterns of bird species richness with spatially explicit models. Proc R Soc B 274:165–174

Rapoport EH (1982) Areography. Geographical strategies of species. Pergamon Press, Oxford, UK

Ricklefs RE (2004) A comprehensive framework for global patterns in biodiversity. Ecol Lett 7:1–15

Stevens GC (1989) The latitudinal gradient in geographical range: how so many species coexist in the tropics. Am Nat 133:240–256

Sunday JM, Bates AE, Kearney MR (2014) Thermal-safety margins and the necessity of thermoregulatory behavior across latitude and elevation. P Natl Acad Sci USA 111:5619–5615

Trakimas G, Whittaker RJ, Borregaard MK (2016) Do biological traits drive geographical patterns in European amphibians? Global Ecol Biogeogr 25:1228–1238

Vági B, Végvári Z, Liker A et al (2019) Parental care and the evolution of terrestriality in frogs. P Roy Soc B-Biol Sci 286:20182737

VanDerWal J, Murphy HT, Lovett-Doust J (2008) Three-dimensional mid-domain predictions: geometric constraints in North American amphibian, bird, mammal and tree species richness patterns. Ecography 31:435–449

Vasconcelos TS, dos Santos TG, Haddad CHB, Rossa-Feres DC (2010) Climatic variables and altitude as predictors of anuran species richness and number of reproductive modes in Brazil. J Trop Ecol 26:423–432

Vasconcelos TS, Rodríguez MA, Hawkins BA (2012) Species distribution modelling as a macroecological tool: a case study using New World amphibians. Ecography 35:539–548

Villalobos F, Dobrovolski R, Provete DB, Gouveia SF (2013) Is rich and rare the common share? Describing biodiversity patterns to inform conservation practices for South American anurans. PLoS ONE 8:e56073

Vitt LJ, Caldwell JP (2009) Herpetology: an introductory biology of amphibians and reptiles. Elsevier, Burlington, MA

Werner EE, Skelly DK, Relyea RA, Yurewicz KL (2007) Amphibian species richness across environmental gradients. Oikos 116:1697–1712

Whitton FJS, Purvis A, Orme CDL, Olalla-Tárraga MÁ (2012) Understanding global patterns in amphibian geographic range size: does Rapoport rule? Global Ecol Biogeogr 21:179–190

Wiens JJ, Graham CH (2005) Niche conservatism: integrating evolution, ecology, and conservation biology. Annual Rev Ecol Evol S 36:519–539

Yee DA, Juliano SA (2007) Abundance matters: a field experiment testing the more individuals hypothesis for richness–productivity relationships. Oecologia 153:153–162

Chapter 4
Spatial Distribution of Phylogenetic Diversity of South American Anurans

Abstract Understanding the spatial distributions of phylogenetic diversity is an opportunity to support policy-makers and designing conservation strategies in megadiverse regions. Here, we mapped the spatial distribution of Faith's phylogenetic diversity (PD) and phylogenetic endemism (PE) of amphibian species distributed across South America. Although we found areas of high species richness (SR) correlated with areas of high PD or PE, there are regions with much more PD/PE or much less PD/PE than expected given the SR. Using a phylogenetic approach, we found that the factors regulating amphibian biodiversity involve a complex interplay of evolutionary and biogeographical processes in different regions of South America. These results might help supporting conservation planning for this threatened vertebrate group.

Keywords Amphibians · Coldspot · Conservation biogeography · Evolutionary history · Faith diversity · Phylogenetic endemism

4.1 Introduction

Conservation science is concerned with helping preserve the Earth's biodiversity. However, biodiversity is a multifaceted concept (Magurran and McGill 2011), and most conservation approaches are based only on taxonomic diversity, which assumes all species are equally different from each other, leading to an incomplete view of how biodiversity is distributed (Forest et al. 2007; Devictor et al. 2010; Vellend et al. 2011). The phylogenetic diversity, as proposed by Vane-Wright et al. (1991), intends to quantify the amount of evolutionary history contained in a given community by calculating pair-wise species distance from a phylogenetic tree. This biodiversity metric integrates information on the phylogenetic positions of species as a legacy of evolutionary processes and is highly useful for assisting conservation assessments (Webb et al. 2002; Mouquet et al. 2012; Diniz-Filho et al. 2013; Winter et al. 2013; Veron et al. 2019). The integration of phylogenetic diversity into

© Springer Nature Switzerland AG 2019   99
T. S. Vasconcelos et al., *Biogeographic Patterns of South American Anurans*,
https://doi.org/10.1007/978-3-030-26296-9_4

conservation practices is relevant because many species that are both evolutionarily distinct and globally endangered may not be benefited from existing protected areas (Forest et al. 2007; Isaac et al. 2007; Redding et al. 2010). For instance, Forest et al. (2007) showed that the best set of protected areas in the Cape floristic province of South Africa differs when the criteria for selection used was phylogenetic diversity or species richness. Rosauer et al. (2017) found that the selection of global land for conservation priority based on phylogenetic diversity has the potential to capture up to an additional 5943 Myr of distinct mammal evolutionary history compared to the species richness scenario within the same total area. Thus, spatial conservation planning can be improved by identifying regions with high levels of phylogenetic information in addition to species-rich areas, which are not necessarily congruent (i.e., Forest et al. 2007; Devictor et al. 2010; Winter et al. 2013; Misher et al. 2014; Rosauer et al. 2017; but see Rapacciuolo et al. 2018 for different view).

Amphibians are the most threatened vertebrate group globally, with about one-third of species being currently threatened with extinction and half of them in decline (Catenazzi 2015). The main threats include fungal diseases, habitat destruction and alteration, and climate change (Catenazzi 2015). Paradoxically, amphibians also have one of the largest Linnean (i.e., species still unknown), Wallacean (i.e., incomplete knowledge of geographical distributions), and Darwinian (i.e., the lack of relevant phylogenetic information for most organisms) shortfalls (Ficetola et al. 2014; Hortal et al. 2015). For instance, in the International Union for Conservation of Nature (IUCN) Red List of amphibians, 1294 species (22.5%) are not evaluated due to data deficiency, but several can be identified as evolutionarily highly distinct (Isaac et al. 2012). Fritz and Rahbek (2012) found that areas around the world with unusually low phylogenetic diversity of amphibians, as expected based on their species richness, were mostly on islands, indicating large radiations of few lineages that have successfully colonized these archipelagos. Thus, identifying and understanding the processes influencing the spatial distribution of evolutionary history of amphibians is mandatory, particularly in megadiverse regions.

In face of the current amphibian extinction crisis (Wake and Vredenburg, 2008), analyzing the spatial distribution of phylogenetic diversity is an opportunity to assist policy-makers and conservation plans (Rosauer et al. 2017; Veron et al. 2019). Therefore, we mapped the spatial distribution of alternative phylogenetic diversity metrics of anuran species across South America. We also made use of null model analysis to compare whether the phylogenetic diversity of each grid cell is higher/lower than expected to its respective value of species richness. All in all, the spatial distribution of anuran phylogenetic diversity found herein will be used to assist in the final spatial conservation planning in Chap. 7.

4.2   Material and Methods

4.2.1   Anuran Species Distribution

We are using the same anuran species distribution dataset described in Chap. 1 of this book. In summary, we downloaded extent-of-occurrence maps for all anuran species recorded in South America from the IUCN *version* 2016 and created the species ranges for those species not included in the IUCN database. Then, we overlaid the range maps onto a grid with 1649 cells at 100 × 100 km to generate a presence-absence matrix in South America.

4.2.2   Amphibian Phylogeny

We used the time-calibrated tree proposed by Jetz and Pyron (2018), which contains 7238 amphibian species (94% of known extant species). We pruned this tree to include only the species found in South America. However, 181 species were not included in the time-calibrated tree. We decided not to include these species because (*i*) polytomies under-sample branch length differences among species and (*ii*) all species out of the tree belong to genera already represented in the phylogeny. Therefore, the metrics of phylogenetic diversity were calculated considering 2421 species (Fig. 4.1).

Fig. 4.1 Circular time-calibrated tree displaying 2421 anuran species. Green rings indicate time intervals of 25 Mya. Phylogeny proposed by Jetz and Pyron (2018)

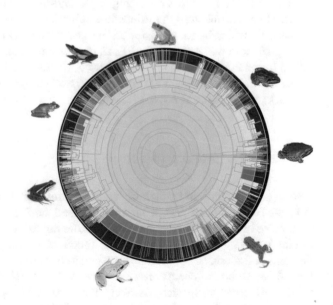

4.2.3   Phylogenetic Indexes

There are many metrics of phylogenetic diversity designed to quantify different aspects of evolutionary relationships (Pavoine et al. 2009; Vellend et al. 2011; Tucker et al. 2016). Thus, choosing the right metric is not an easy task. Tucker et al. (2016) evaluated and classified 70 of them based on their mathematical formulae within three dimensions: (i) "richness," representing the sum of accumulated phylogenetic differences among taxa; (ii) "divergence," representing the average phylogenetic difference between taxa in an assemblage; or (iii) "regularity," representing how regular the phylogenetic differences between taxa in an assemblage are. Here, we calculated Faith's phylogenetic diversity (Faith 1992) and phylogenetic endemism (Rosauer et al. 2009) that represent metrics of richness dimension (Tucker et al. 2016) and are usually used to assist in conservation decisions. Faith's phylogenetic diversity (PD) is the sum of branch lengths connecting all species in an assemblage (Faith 1992). Phylogenetic endemism (PE) combines the widely used PD and weighted endemism measures to identify areas where substantial components of phylogenetic diversity are restricted (Rosauer et al. 2009). Thus, PE applies a concept of endemism based on lineages rather than species, recognizing, for example, that a narrow range endemic species may not represent a large degree of endemism if it is closely related to a widespread species (Rosauer et al. 2009).

Phylogenetic diversity metrics are usually correlated with species richness (Vellend et al. 2011; Tucker et al. 2016). Thus, to remove the influence of species richness (SR) from PD and PE, we calculated the standardized effect size (SES) based on null models (Kembel 2009; Swenson 2014). First, we generated random values of PD and PE values for each grid by shuffling the tips of the phylogeny 999 times. Then, the observed PD values was subtracted from the average of simulated values and divided by standard deviation of simulated PD to calculate SES.PD. The same procedure was repeated for PE. Grid cells containing negative values of SES reflect lower PD and PE than expected for SR, while grid cells containing positive values of SES reflect greater PD and PE than expected for SR. For normally distributed data, significance at $P < 0.05$ is equivalent to a $1.96 > SES < -1.96$.

4.3   Results and Discussion

Two regions recognized by their high annual rainfall and altitudinal gradients – the Amazon basin and its adjacent part of Andes and the Serra do Mar and Mantiqueira mountains in southeastern Brazil – contains the highest values of SR, PD, and PE in South America (Fig. 4.2). In contrast, much of Caatinga, Cerrado, and Chaco domains and southern South America regions supported the lowest SR, PD, and PE (Fig. 4.2). Mountain chains are reservoirs of biodiversity, harboring around one-third of all terrestrial species (Antonelli et al. 2018). This phenomenon is driven by two basic mechanisms: climatic barriers to dispersal (Janzen 1967; Polato et al.

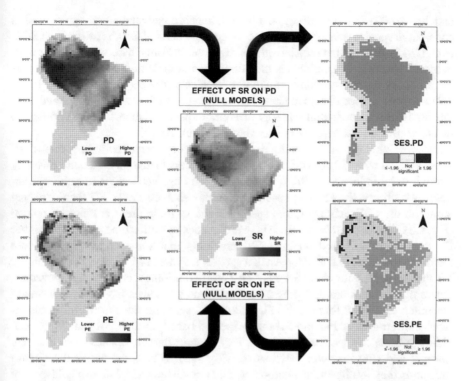

Fig. 4.2 Spatial distribution of amphibian species richness (SR), Faith's phylogenetic diversity (PD), phylogenetic endemism (PE), standardized effect size of PD (SES.PD), and standardized effect size of PE (SES.PE) along South America. Values of SES ≤ −1.96 indicate grid cells that contain significantly less PD and PE than expected for the number of SR present. Values of SES ≥ 1.96 indicate grid cells that contain significantly more PD and PE than expected for the number of SR present

2018) and allopatric speciation due to formation of mountains. Tropical mountain chains are expected to favor speciation more than temperate mountains, due to narrow climatic niche of tropical species, which tolerate less variation in climate than temperate species (Janzen 1967; Polato et al. 2018). This mechanism hampers dispersal and promotes in situ speciation. Therefore, it follows that mountainous regions are also centers of endemism, working as a pump of speciation mostly due to allopatric speciation as a result of geological and orogenetic history that presumably created dispersal barriers to organisms (Graham et al. 2014, 2018; Merckx et al. 2015; Badgley et al. 2017). Non-Andean mountains in South America are especially important as centers of endemism for several taxa (Guedes et al. 2019). Among them, the tops of the Serra do Mar and Mantiqueira share a number of plant clades with the Andes, due to similar past climatic conditions since the Pleistocene/ Holocene. Those same regions stood out in our analysis as having high PE, which concur with the above-cited processes. Additionally, climatic gradients have also presumed to influence the geographic distribution of anurans. For instance, Buckley and Jetz (2007) found a strong positive relationship between annual precipitation and anuran species richness worldwide. In specific parts of South America, da Silva et al. (2012) showed that moister sites in the Atlantic Forest harbored a greater phylogenetic diversity of amphibians than drier sites, indicating that precipitation gradients act as barriers to some amphibian species with specific life-history traits. Taken together, these results indicate strong signatures of mountain chains and rainfall distributions on spatial patterns of anuran species richness and phylogenetic diversity in South America (see also Antonelli et al. 2018).

Because phylogenetic diversity metrics are usually correlated with species richness (Vellend et al. 2011; Tucker et al. 2016), we evaluated phylogenetic diversity removing the influence of SR by comparing the observed values to null expectations. We observed that most species-rich areas showed lower PD than expected based on their SR indicating that species found in these areas are closely related (Fig. 4.2). On the other hand, areas in part of Andes in Amazon basin showed higher PE than expected based on their SR indicating that some anuran lineages are restricted to this region. We suggest that species-rich areas with lower PD and higher PE than expected for their SR may be both evolutionary cradles and museums, where old lineages are still present and many new lineages have been constantly generated (Fritz and Rahbek 2012). Middle and southern portion of Andes mountains showed greater PD and PE than expected based on their SR indicating that, although these are species-poor areas, they support anuran species from different lineages and with restricted distributions (Fig. 4.2). Therefore, our results support the notion that South America harbors widely known hotspots, but also coldspots of species richness and phylogenetic diversity. While the Cerrado and the Atlantic Forest have received a great deal of attention from the perspective of systematic conservation planning in the past decades (e.g., Villalobos et al. 2013; Loyola et al. 2014), coldspots did not have the same fate. However, although being poor in anuran species, these areas are certainly of great conservation value.

Acknowledgments The authors have been continuously supported by research grants and/or fellowships from the Fundação de Amparo à Pesquisa do Estado de São Paulo (FAPESP 2011/18510-0; 2013/50714-0; 2016/13949-7), Conselho Nacional de Desenvolvimento Científico e Tecnológico (CNPq 2037/2014-9; 431012/2016-4; 308687/2016-17; 114613/2018-4), and University Research and Scientific Production Support Program of the Goias State University (PROBIP/UEG). Prof. Dr. Victor Satoru Saito (UFSCar) read critically the first version of this manuscript and provided insightful comments that improved it.

References

Antonelli A, Kissling WD, Flantua SGA et al (2018) Geological and climatic influences on mountain biodiversity. Nat Geosci 11:718–725
Badgley C, Smiley TM, Terry R et al (2017) Biodiversity and topographic complexity: modern and geohistorical perspectives. Trends Ecol Evol 32:211–226
Buckley LB, Jetz W (2007) Environmental and historical constraints on global patterns of amphibian richness. Proc R Soc Lond 274:1167–1173

Catenazzi A (2015) State of the World's amphibians. Annu Rev Environ Resour 40:91–119

da Silva FR, Almeida-Neto M, Prado VHM et al (2012) Humidity levels drive reproductive modes and phylogenetic diversity of amphibians in the Brazilian Atlantic Forest. J Biogeogr 39:1720–1732

Devictor V, Mouillot D, Meynard C et al (2010) Spatial mismatch and congruence between taxonomic, phylogenetic and functional diversity: the need for integrative conservation strategies in a changing world. Ecol Lett 13:1030–1040

Diniz-Filho JA, Loyola RD, Raia P et al (2013) Darwinian shortfalls in biodiversity conservation. Trends Ecol Evol 28:689–695

Faith DP (1992) Conservation evaluation and phylogenetic diversity. Biol Conserv 61:1–10

Ficetola GF, Rondinini C, Bonardi A et al (2014) An evaluation of the robustness of global amphibian range maps. J Biogeogr 41:211–221

Forest F, Grenyer R, Rouget M et al (2007) Preserving the evolutionary potential of floras in biodiversity hotspots. Nature 445:757–760

Fritz SA, Rahbek C (2012) Global patterns of amphibian phylogenetic diversity. J Biogeogr 39:1373–1382

Graham CH, Carnaval AC, Cadena CD et al (2014) The origin and maintenance of montane diversity: integrating evolutionary and ecological processes. Ecography 37:711–719

Graham CH, Parra M, Mora A et al (2018) The interplay between geological history and ecology in mountains. In: Hoorn C, Perrigo A, Antonelli A (eds) Mountains, climate and biodiversity. Wiley Blackwell, Hoboken, p 231

Guedes TB, Azevedo JAR, Bacon CD et al (2019) Diversity, endemism, and evolutionary history of montane biotas outside the Andean region. In: Rull V, Carnaval A (eds) Neotropical Diversification. Springer, New York

Hortal J, de Bello F, Diniz-Filho JAF et al (2015) Seven shortfalls that beset large-scale knowledge on biodiversity. Annu Rev Ecol Evol Syst 46:523–549

Isaac NJB, Turvey ST, Collen B et al (2007) Mammals on the EDGE: conservation priorities based on threat and phylogeny. PLoS One 2:e296

Isaac NJB, Redding DW, Meredith HM (2012) Phylogenetically-Informed Priorities for Amphibian Conservation. PLoS One 7(8):e43912

IUCN (2016) IUCN red list of threatened species. http://www.iucnredlist.org/technical-documents/spatial-data. Accessed 22 Mar 2019

Janzen DH (1967) Why mountain passes are higher in the tropics. Am Nat 101:233–249

Jetz W, Pyron RA (2018) The interplay of past diversification and evolutionary isolation with present imperilment across the amphibian tree of life. Nat. Ecol. Evol. 2:850–858

Kembel SW (2009) Disentangling niche and neutral influences on community assembly: assessing the performance of community phylogenetic structure test. Ecol Lett 12:949–960

Loyola RD, Lemes P, Brum FT et al (2014) Clade-specific consequences of climate change to amphibians in Atlantic Forest protected areas. Ecography 37:65–72

Magurran AE, McGill BJ (2011) Biological diversity Frontiers in measurement and assessment. Oxford University Press, Oxford

Merckx VSFT, Hendricks KP, Beentjes KK et al (2015) Evolution of endemism on a young tropical mountain. Nature 524:347–350

Misher BD, Knerr N, González-Orozco CE et al (2014) Phylogenetic measures of biodiversity and neo- and paleo-endemism in Australian Acacia. Nat Commun 5:4473

Mouquet N, Devictor V, Meynard CN et al (2012) Ecophylogenetics: advances and perspectives. Biol Rev 87:769–785

Pavoine S, Love MS, Bonsall MB (2009) Hierarchical partitioning of evolutionary and ecological patterns in the organization of phylogenetically-structured species assemblages: application to rockfish (genus:Sebastes) in the Southern California Bight. Ecol Lett 12:898–908

Polato NR, Gill BA, Shah AA et al (2018) Narrow thermal tolerance and low dispersal drive higher speciation in tropical mountains. Proc. Nat. Acad. Sci. 115:12471–12476

Rapacciuolo G, Graham CH, Marin J et al (2018) Species diversity as a surrogate for conservation of phylogenetic and functional diversity in terrestrial vertebrates across the Americas. Nat Ecol Evol 3:53–61

Redding DW, DeWolff CV, Mooers AO (2010) Evolutionary distinctiveness, threat status, and ecological oddity in primates. Conserv Biol 24:1052–1058

Rosauer DF, Laffan SW, Crips MD et al (2009) Phylogenetic endemism: a new approach for identifying geographical concentrations of evolutionary history. Mol Ecol 18:4061–4072

Rosauer DF, Pollock LJ, Linke S (2017) Phylogenetically informed spatial planning is required to conserve the mammalian tree of life. Proc R Soc Lond 284:20170627

Swenson NG (2014) Functional and Phylogenetic Ecology in R. Springer, New York

Tucker CM, Cadotte MW, Carvalho SB et al (2016) A guide to phylogenetic metrics for conservation, community ecology and macroecology. Biol Rev 92:698–715

Vane-Wright RI, Humphries CJ, Williams PH (1991) What to protect? Systematics and the agony of choice. Biol Conserv 55:235–254

Vellend M, Cornwell WK, Magnuson-Ford K (2011) Measuring phylogenetic biodiversity. In: Magurran AE, McGill BJ (eds) Biological diversity Frontiers in measurement and assessment. Oxford University Press, Oxford

Veron S, Saito V, Padilla-García N et al (2019) The use of phylogenetic diversity in conservation biology and community ecology: a common base but different approaches. Q Rev Biol 94:123–148

Villalobos F, Dobrovolski R, Provete DB et al (2013) Is rich and rare the common share? Describing biodiversity patterns to inform conservation practices for South America anurans. PLoS One 8:e56073

Wake DV, Vredenburg VT (2008) Are we in the midst of the sixth mass extinction? A view from the world of amphibians. Proc Nat Acad Sci 105:1146–11473

Webb CO, Ackerly DD, McPeek MA et al (2002) Phylogenies and community ecology. Ann Rev Ecol Evol Syst 33:475–505

Winter M, Devictor V, Schweiger O (2013) Phylogenetic diversity and nature conservation: where are we? Trends Ecol Evol 28:199–204

Chapter 5
Geographical Patterns of Functional Diversity of South American Anurans

Abstract A critical step in designing efficient spatial prioritization plans is to find the best network of sites that preserve multiple biodiversity facets. Conservation biogeography has increasingly been using functional diversity (FD) as an alternative metric to describe how trait diversity is distributed throughout space. FD can be promptly related to the function played by species in a community, better than taxonomic diversity. Here, we mapped multiple dimensions of functional diversity of South American anurans that describe functional richness (FRich), evenness (FEve), dispersion (FDis), and rarity (Frar) at the regional scale, as well as geographical restrictiveness and body size. FRich was higher in the Amazon basin, Guianas, Tropical Andes, and the central portion of the Atlantic Forest, whereas lower values appear in Patagonia and the Atacama Desert. FEve and FDis were homogeneously distributed throughout the continent, with lowest values in southern Patagonia, Pacific slope of the southern Tropical Andes, and Atacama Desert, while the highest values were found in the Atlantic slope of the southern Tropical Andes. Patterns of functional distinctiveness and uniqueness were similar and highlighted northwestern Argentina and areas with many endemic and functionally unique species. Frar at the regional scale highlighted the Andes, Patagonia, and the Atlantic Forest.

Keywords Body size · Functional biogeography · Functional rarity · Life-history traits · Conservation biogeography · South America

5.1 Introduction

Functional traits are "morpho-physiophenological traits which impact fitness indirectly via their effects on growth, reproduction and survival, the three components of individual performance" (*sensu* Violle et al. 2007). Functional traits have long been used to understand community assembly, especially of plants (see McGill et al. 2006 and Mason and de Bello 2013 for reviews). Among the advantages of using traits instead of species names is their ability to discriminate community assembly processes, such as competition and environmental filters, since traits can be directly linked to resource use and environmental requirements (McGill et al. 2006; Violle et al. 2007), both aspects of the contemporary ecological niche concept (Chase and Leibold 2003). Because of research tradition, functional traits have been usually employed in community-scale studies, but less so at continental and global scales. This trend has changed with the emergence of functional biogeography (Violle et al. 2014).

Functional biogeography aims to understand broad-scale spatial patterns of functional diversity and its relationship with environmental gradients (Violle et al. 2014). This new approach invites a change in perspective from the traditional ways in which we studied species distribution (see Lomolino et al. 2016) using species as units. Instead, functional biogeography analyzes how functional traits vary throughout space and their drivers. It can be used both as an empirical assessment of how traits vary at broad scales and its causes and also as a tool to select and prioritize areas for reserves (e.g., Ouchi-Melo et al. 2018). This approach is also a complementary way to analyze how environmental gradients or filters change trait composition in communities, by looking at them at a broad spatial scale (see Newbold et al. 2012).

While there are standardized protocols for measuring functional traits of spermatophytes (Cornelissen et al. 2003; Pérez-Harguindeguy et al. 2013), phytoplankton (Reynolds et al. 2002; Litchman and Klausmeier 2008; Kruk et al. 2010), zooplankton (Litchman et al. 2013), and terrestrial invertebrates (Moretti et al. 2017), no such protocol exists for amphibians. The lack of consensus on what traits to measure has impaired the progress of this research agenda for amphibians. The emergence of large-scale trait databases for vertebrate clades, such as birds, mammals (Jones et al. 2009; Wilman et al. 2014), squamates (Myhrvold et al. 2015), and fishes (Froese and Pauly 2000), ignited the research on the broad-scale spatial distribution of functional diversity of several vertebrate groups (e.g., Oliveira et al. 2016). However, amphibians were lagging behind until the recent publication of amphiBIO (Oliveira et al. 2017), which provided trait data for all known amphibian species up to 2011. Previously, trait data were available for European amphibians (Trochet et al. 2014) and for some South American species in the supplementary material of a few articles (e.g., Ernst et al. 2012).

The choice of which traits to use in functional diversity studies may affect the study conclusions (see Tsianou and Kallimanis 2015, 2018). Ideally, the choice should be driven by the questions (Lefcheck et al. 2014; Zhu et al. 2017) and spatial scale (Violle et al. 2007; Tsianou and Kallimanis 2015). Specifically, it should be recognized the difference between effect and response traits (Violle et al. 2007), which capture different niche dimensions (Chase and Leibold 2003; Rosado et al. 2016). Effect traits are related to the impact species cause in the community or ecosystem, whereas response traits are involved in the way species respond to environmental gradients (Violle et al. 2007). For our goal in this chapter, we chose response traits of South American amphibians that are known to vary across broad spatial gradients, such as body size, habitat (vertical distribution), and development mode that are available in amphiBIO (Oliveira et al. 2017). But before delving into the analysis and its results, it would be timely to review the brief history of functional diversity studies on amphibians worldwide to detect trends and main findings.

5.1.1 A Short History of the Functional Diversity Research Agenda as Applied to Amphibians

Functional diversity has been used to understand community structure of both adults and tadpoles. This approach has proven especially important in recognizing the effects of land use change and habitat alteration on amphibian diversity beyond species names. Another usual topic is broad-scale pattern of spatial distribution of functional diversity, but the studies addressing this topic have only been developed in Europe. Below, we review all papers published to date using functional diversity with amphibians.

Early papers using functional traits of adult amphibians to test community assembly focused on the impacts of logging and environmental disturbances in northern South America and Western Africa (Ernst et al. 2006, 2007). Those two papers calculated Petchey's functional diversity (FD) with five traits of both adults and tadpoles and found that FD was higher in primary forest than in logged areas, although the absolute FD value was different between the two continents. Trimble and van Aarde (2014) later analyzed how functional group richness and abundance of the herpetofauna respond to a gradient of land use, from primary forests to *Eucalyptus* and sugarcane plantations in South Africa. They used four traits for anurans including body size and others involved in habitat selection to define four functional groups. They found that the abundance of functional groups varied along the land use gradient, while functional group richness decreased from forest to cultivation. Gallmetzer and Schulze (2015) analyzed how FD of communities of amphibians and reptiles varied along a gradient of land use from forest interior, forest edge to oil palm plantation in Costa Rica. They used five traits for amphibians, activity time, body size, oviposition site, egg clutch size, and microhabitat, which are used to calculate four FD metrics, functional richness (FRich), functional evenness (FEve), functional dispersion (FDis), and functional divergence (FDiv). They found that all FD metrics, but FEve, differed among the three types of land use, showing lower values for oil palm sites. Additionally, species richness was not different among the land use types. Riemann et al. (2017) tested the effects of habitat fragmentation and land use change on Malagasy frogs using 12 traits related to adult morphology, habitat use, and daily activity. They calculated standardized functional richness (SES.FD) and functional beta diversity (unweighted UniFrac). They found that land use change altered functional beta diversity, but not richness, suggesting that altered habitats contain different, but not necessarily less, functions. Lipinski et al. (2018) analyzed how taxonomic, functional, and phylogenetic diversity of adult anurans varied along an agricultural-forest gradient in Southern Brazil. They sampled 38 ponds and used 7 traits, including body size, habitat, reproductive mode, reproductive phenology, and morphometrical measures. They related Rao's QE to environmental gradients and found that functional diversity and redundancy were not related to distance to the nearest border, but apparently specialized reproductive modes and morphometrical measurements were affected. The summary presented in this paragraph reinforces that the choice of metrics is not straightforward and can alter study's conclusions. We will discuss this matter further below.

To understand anuran community assembly in the Chaco, Lescano et al. (2018) correlated four morphometric measurements of adult amphibians and two measurement of tadpoles, which they refer to as "functional traits," with percentage of canopy cover, pond area and depth (as a proxy for hydroperiod), and herbaceous vegetation around pond margins. Authors found that temporary ponds with higher canopy cover had lower functional diversity. Several papers have used functional traits to investigate the influence of environmental variables on tadpole community structure. For example, Both et al. (2011) found that tadpole guild composition was influenced by pond hydroperiod and depth, with ponds at distinct points of the gradient sustaining tadpole communities with different guild composition in grasslands of Southern Brazil. Queiroz et al. (2015) tested the influence of pond depth on tadpole functional dispersion (FDis) in an agricultural landscape in Southeastern Brazil, finding a unimodal relationship. Strauß et al. (2010) found that tadpole communities of Madagascar forest streams had lower Petchey's FD than expected given their species richness, suggesting that richer streams had a functionally redundant set of species; they used eight traits related to morphology and habitat use. In a follow-up study, the same authors tested the influence of seasonal changes in environmental variables on tadpole functional diversity (Strauß et al. 2016). They found that Petchey's FD decreased less than expected by chance from the wet to the dry season and in a nonrandom way; communities were functionally redundant during the dry season.

Broadscale studies investigated either trait-environment relationship across continents or the spatial distribution of functional diversity. For example, Ernst et al. (2012) used two methods to relate traits to environmental variables and found that trait-environment relationships changed depending on the region at three tropical forests: northern South America, Borneo Rain Forest, and Southeast Africa. Their results suggest that this idiosyncrasy is driven by the phylogenetic composition of the regional species pool, resulting in contrasting species response to environment at different regions. Nonetheless, a common trend was that traits related to broad habitat selection and daily activity were conserved across regions, suggesting that they may represent old adaptations. Trakimas et al. (2016) analyzed the effects of four life-history traits on range size of European amphibians to test the prediction that northern species have larger range sizes than southern ones, due to life-history traits that favor dispersal and colonization of new habitats. They confirmed their hypothesis, but the effects of specific traits were different for anurans and salamanders. Nowakowski et al. (2017) tested if species traits were a better predictor of amphibian sensitivity to habitat modification than the expert-based data from IUCN. They used published data on species abundance in preserved and adjacent disturbed habitats across five continents. They found that species with high sensitivity were also threatened with extinction and had declining population trends, whereas those tolerant were also likely to be invasive outside of their range. Species with larger range size and with pond-dwelling larvae had higher tolerance to habitat modification.

So far, there are only two studies that mapped amphibian functional diversity across an entire continent. Tsianou and Kallimanis (2015) calculated six different

FD metrics and tested if distinct sets of traits (morphological, reproductive, and habitat related) recover similar spatial patterns of amphibian FD across Europe. They found little congruence in the hotspots of FD identified with different types of traits for the same metric, suggesting that the choice of traits is key to functional diversity studies even at a broad spatial scale. A similar result for inferring community assembly processes was found by the same authors using null-model analysis (Tsianou and Kallimanis 2018).

Finally, in this chapter we provide a functional biogeography perspective on South American amphibians. Specifically, we mapped alternative dimensions of functional diversity using distance-based, multivariate α functional diversity indices. To the best of our knowledge, this is a first attempt to provide a broadscale map of ecologically relevant traits for amphibians in South America. Since we used traits similar to other studies with tetrapods at the continental or global scales using comparable datasets (e.g., Oliveira et al. 2016), our results can be useful to understand and compare the mechanisms influencing trait distributional patterns at a biogeographical scale of different groups. These results can inform further conservation biogeography analysis aiming to design systematic spatial conservation planning that improve protected area networks in South America.

5.2   Material and Methods

5.2.1   Trait Data

Here, we used a subset of the traits available in amphiBIO that are known to vary at a broad spatial scale (i.e., Grinnellian niche; Rosado et al. 2016) and are correlated with environmental gradients, such as body size (i.e., Bergmann's rule; Slavenko and Meiri 2015; Amado et al. 2019), developmental mode of larvae (da Silva et al. 2012; Gomez-Mestre et al. 2012; Müller et al. 2013), and habitat (Table 5.1). To calculate trait dissimilarity matrix, we used the Gower index that allows the incorporation of mixed data types (Pavoine et al. 2009; Legendre and Legendre 2012). We also checked if the distance matrix had Euclidean properties.

However, amphiBIO has only 30% of data completeness and does not include all anuran species known to occur currently in South America included in our dataset (see Chap. 1). This fact is an example of the Raunkiæran shortfall (Hortal et al. 2015, see also Rosado et al. 2016) common in megadiverse regions. Specifically, of the 2623 species in our dataset, amphiBIO had trait data for only 2202 species (83.9%). However, given that 22.25% (490 species) of those 2202 had missing data for body size, which is a key trait, we updated the dataset by searching for body size data in the literature (e.g., Amado et al. 2019; Supplementary Material). We corrected for nomenclature changes and further removed synonyms, nomina nuda, and nomina dubia established between 2017 and 2018, reducing our dataset to a final 2586 species (see Supplementary Material). Yet, maximum body size was not available for 34 species, representing 0.76% of missing data. For further analysis, we log2 transformed body size to improve index accuracy (see Májeková et al. 2016).

Table 5.1 Traits used to calculate functional diversity, taken from the amphiBIO database (Oliveira et al. 2017) and from the literature (Amado et al. 2019; Supplementary Material)

Trait	Data type	Type of trait	Ecological interpretation	References
Maximum body size	Continuous	Effect/response	Correlated with several life-history characteristics, including fitness	Peters (1986)
Type of habitat	Categorical multi-choice, coded as dummy	Response	Provides a measure of space use and vertical distribution. Microhabitat used most of the time by species: terrestrial, aquatic, arboreal, and fossorial	Lion et al. (2019), Oliveira and Scheffers (2018)
Larval developmental mode	Binary	Response	Evidence suggest that direct development is related to humidity levels. Viviparous, larvae, direct development	Da Silva et al. (2012), Gomez-Mestre et al. (2012)

5.2.2   Data Analysis

The choice of which traits and functional index to use is very contentious (Schleuter et al. 2010; Mason et al. 2013) and depends on the study goals. Several critiques have been posed recently on the poor performance of some indexes in identifying community assembly processes with simulated data (see McPherson et al. 2018). We avoided the use of indexes that are monotonically related to species richness, in order to distinguish between these two facets of biodiversity. Thus, we calculated three commonly used functional diversity indexes that describe independent aspects of functional space (Mason et al. 2005): functional richness (FRich; Villéger et al. 2008), unweighted functional evenness (FEve), and functional dispersion (FDis; Laliberté and Legendre 2010). We did not use functional divergence (FDiv) due to recent concerns about the index (see Kuebbing et al. 2017).

FRich describes the total range of functional trait variability of a community (Legras et al. 2018) by measuring the smallest convex hull containing all species. As such, it is more influenced by the vertices of the convex hull, i.e., species with extreme values. Then, this index provides information on shifts in functional volume along environmental gradients and is suitable for our purposes because we have more species than traits. Accordingly, we used the unweighted version of FEve that does not take into account species abundance (Legras and Gaertner 2018), because the presence-absence matrix contains only species incidence in grid cells. Calculated only with species incidence, FEve is a measure of regularity of functional distances and does not suffer from the critical issues pointed out recently (see Legras and Gaertner 2018). The unweighted version of FDis measures the distance between an individual species and the centroid of all species of the community (Laliberté and Legendre 2010); it is not affected by species richness and outliers. All metrics are contingent on the species pool under consideration, in our case South America. As such, it will change if restricted to smaller specific regions or biomes.

Additionally, we also calculated recently proposed indexes of functional rarity for each grid cell: mean functional uniqueness, mean functional distinctiveness, and mean geographical restrictiveness (Violle et al. 2017). Functional distinctiveness (D_i) is the average functional distance from a species to all others in the grid; it varies from 0 (when a given species is functionally close – redundant – to many others) to 1 (when all species in the grid have a maximum distance to the focal species). Therefore, it is equivalent to mean pairwise distance (MPD; see Chap. 4). Functional uniqueness (U_i) represents how "isolated" a species in the grid is; it is the functional distance to the nearest neighbor of a given focal species. Therefore, it is equivalent to Mean Nearest Taxon Distance (MNTD; see Chap. 4). The main difference between D_i and U_i is the scale at which they are measured: uniqueness is a regional-scale measure, whereas distinctiveness is a local-scale measure. Geographical restrictiveness (R_i) is related to the extent of a species; it varies from 1 (when the species occurs in only one grid cell, i.e., restricted) to 0 (when the species is present at all grid cells). Given the extent of our study, we were interested in visualizing geographical patterns of rarity at the regional scale, i.e., taking into account functional uniqueness and restrictiveness. Thus, following Violle et al. (2017: 362), we calculated functional rarity at the regional scale by adding up U_i and R_i. To calculate the mean of all indices, we used the function funtcomp that implements a community-weighted mean (Garnier et al. 2004). Analyses were conducted in the R packages FD (Laliberté et al. 2014) and funrar (Grenié et al. 2017). Indices were then mapped with ggplot2 (Wickham 2016) and sf (Pebesma 2018).

5.3   Results and Discussion

Despite having used only three functional traits, our analysis recovered interesting geographical patterns. The Tropical Andes was a region with high functional richness (Fig. 5.1a). The Andes is the largest mountain range on Earth and the only one to cross the equator (see Antonelli et al. 2009). It is usually subdivided into three regions based on climate: Tropical Andes from Venezuela to southern Bolivia, Dry Andes from the Atacama Desert to 35° S, and Wet Andes from 35° S to 56° S. Several studies have pointed out that the Andes uplift has played a key role on the origin of biodiversity of South America (e.g., Antonelli et al. 2009; Rangel et al. 2018). The Andes uplift during the Neogene took about 40 Myr to complete and altered both geological configuration and climatic patterns, changing the direction of the Amazon River, altering rainfall regimes and consequently humidity levels in different regions (Antonelli et al. 2009; Hoorn et al. 2010, reviewed in Sepulchre et al. 2011; Vonhof and Kaandorp 2011). Our results further suggest that the Andes seems not only an important region of high species richness (see Chap. 3) but also a diverse region for species phenotypic characteristics (this chapter) and phylogenetic relationships (see Chap. 4). Further studies should test evolutionary and ecological hypothesis to explain the origin of phenotypic diversity in the Andes, including mode of speciation and environmental requirements.

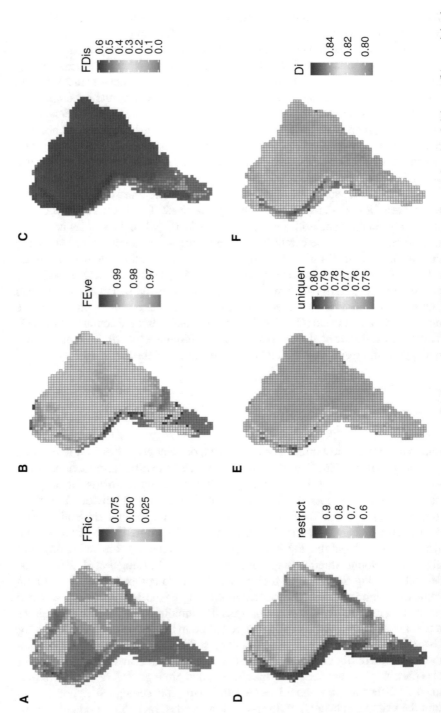

Fig. 5.1 Geographical patterns of functional diversity and functional rarity of anuran amphibians in South America. (**a**) functional richness, (**b**) unweighted functional evenness, (**c**) functional dispersion, (**d**) geographical restrictiveness, (**e**) functional uniqueness, (**f**) functional distinctiveness

The highest values of functional richness are also found in the Amazon basin, Tropical Andes, Guiana Shield, and the Brazilian Atlantic Forest. Recent evidence has shown that Amazonia acted as a pump of species, with high speciation rates for several taxa, which have dispersed to other regions (Antonelli et al. 2018a). The mountain ranges of the Serra do Mar and Serra da Mantiqueira in Southeastern Brazil are known to have high species richness composed essentially of small-ranged species (Villalobos et al. 2013; Vasconcelos et al. 2014). Most of bird and amphibian species in this region are threatened, making it a global biodiversity hotspot (Mittermeier et al. 2005). The biodiversity of this region is comparable to other mountains outside the Andes (Guedes et al. 2019). The high endemism in mountainous regions is usually associated with in situ allopatric speciation (Graham et al. 2014; Badgley et al. 2017; Antonelli et al. 2018b), but there is evidence showing that a good amount of immigration of cold-adapted species also contributes to their high species richness (Merckx et al. 2015). Finally, the high anuran diversity in Amazonia (Wiens et al. 2011) and Tropical Andes can also be explained by colonization time, along with high speciation rates (Hutter et al. 2013, 2017).

Except for the southern Patagonia, Atacama Desert, Dry Andes region, and a small section in the Tropical Andes, which had intermediate values of FEve, values of functional evenness were all high in most of the continent (Fig. 5.1b), with a mean of 0.98 (± 0.0051) in the dataset. Interestingly, the highest values were also associated with the eastern slope of the Dry Andes and northern Patagonia. Therefore, communities in those regions seem to have species with regularly distributed traits, which makes interesting to discover the drivers of low FEve in the regions pointed out above. Regions with low FEve are arid and cold, suggesting that environmental filters play a key role in selecting anuran species that tolerate high seasonality in rainfall and low temperature. This fact may reduce the number and evenness of traits in those communities.

Although having low variation across the continent, functional dispersion (Fig. 5.1c) had a similar pattern to FEve, with the lowest values in the Atacama Desert and southern Patagonia. Functional diversity metrics are dependent on the species pool being analyzed. Using the extent of South America, both FEve and FDis did not seem to provide good insights to understand trait distribution. Therefore, future studies using FEve and FDis should take into account a finer scale, ideally using WWF ecoregions or the biogeographic regions identified in Chap. 6.

The spatial pattern of mean functional uniqueness (Fig. 5.1d) and functional distinctiveness (Fig. 5.1f) are similar and highest in northwestern Argentina, in the provinces of Salta and Jujuy, extending to the Bolivian department of Tarija (Fig. 5.1f). This region is at the Andes foothills and lies at the transition zone between many ecoregions, such as the Dry Chaco, the Southern Andean Yungas, the Central Andean puna, and the Central Andean dry puna (Olson et al. 2001). Except for the Southern Andean Yungas, all those ecoregions are grassland formations, with short vegetation and dry to semiarid. Nevertheless, this region harbors many (micro) endemic species, such as *Pleurodema borellii* and *Rhinella gnustae*, some of them threatened or data deficient, such as *Rhinella rumbolli* and several species of the genus *Telmatobius*, including *T. pinguiculus* and *T. pisanoi*. Our results show

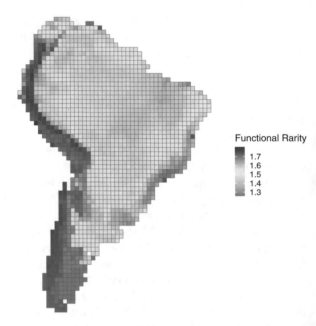

Fig. 5.2 Functional rarity at the regional scale, which consists of the sum of functional uniqueness and geographical restrictiveness. (see Violle et al. 2017:362)

Functional Rarity

1.7
1.6
1.5
1.4
1.3

that, in relation to the species pool of South America, this region harbors many functionally distinct species both at the regional (U_i) and local (D_i) scales. Other areas with relatively high D_i are the Dry and Tropical Andes regions.

Geographical restrictiveness (Fig. 5.1e) was highest in the Andes, Patagonia, Atlantic Forest, and northwestern Argentina. This pattern suggests that species occurring in these regions are mostly small-ranged. A recent paper tested four hypotheses for the origin and maintenance of endemism on Earth (Zuloaga et al. 2019). Authors found that species richness, long-term climate stability, climate seasonality, and spatial heterogeneity in climate, topography, and habitat best explained broad-scale patterns of endemism for amphibian species, but with a larger relative contribution of spatial heterogeneity. Interestingly, their results suggest that species richness and endemism are not predicted by the same set of variables.

Functional rarity (Fig. 5.2) at the regional scale highlights the Andes, northwestern Argentina, Patagonia, and most of the Atlantic Forest as areas with geographically restricted and functionally unique species. In contrast, areas with low functional rarity include the so-called dry diagonal that covers the Brazilian Shield and includes the dry biomes, such as Cerrado and Caatinga.

Amado et al. (2019) were able to assemble data on body size for 2761 species occurring on the New World. In this study, we mapped the mean maximum body size for 2552 species that occur in South America (Fig. 5.3). With relatively more coverage for the region, our results (Fig. 5.3) largely confirm the trends found by Amado et al. (2019): larger species tend to occur in cold regions, such as those in high latitude and high-altitude sites in the Andes. Recent models (Gouveia et al. 2019; Rubalcaba et al. 2019) that combine ecophysical and water economy have found that the distribution of body size in amphibians is better explained by water

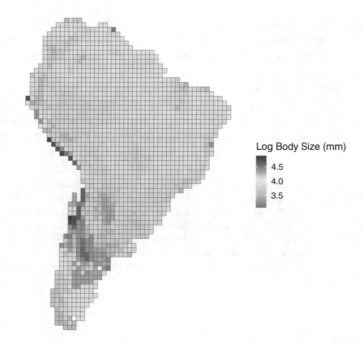

Fig. 5.3 Mean maximum body size in mm (snout-vent length) of anuran amphibians in South America

relationships. Being a considerably larger dataset than those assembled previously, our map for mean body size suggests that Bergmann's rule indeed applies to anurans, as previously suggested (Slavenko and Meiri 2015).

5.3.1   Implications for Spatial Conservation Planning/ Conservation Biogeography

Also, representing novel data, future studies should explore the climatic and geographic drivers of the spatial patterns we demonstrated. For instance, no study has yet investigated the climatic drivers of functional rarity for amphibians at broad spatial scales or if it is affected by land use changes. The results of such study may assist conservation biogeographers in delimiting networks of protected areas.

5.3.2   Recommendations for Future Work

Our intention in this chapter was to describe how different metrics of anuran functional diversity are distributed across South America. We believe that our results remain yet to be fully explored in studies that uncover the drivers of functional diversity in South America. For example, the next studies could use metrics that integrate functional traits to test if hotspots of amphibian diversity, detected using species richness and range size (e.g., Villalobos et al. 2013), are congruent with those of functional diversity.

The choice of traits in functional diversity studies is also a crucial step. Therefore, the geographical patterns recovered here are dependent on the three traits we chose. Through highlighting geographical regions that are already known to support high species richness (see Chap. 3), the patterns could change depending on the traits used.

Acknowledgments　The authors have been continuously supported by research grants and/or fellowships from the Fundação de Amparo à Pesquisa do Estado de São Paulo (FAPESP 2011/18510-0; 2013/50714-0; 2016/13949-7), Conselho Nacional de Desenvolvimento Científico e Tecnológico (CNPq 2037/2014-9; 431012/2016-4; 308687/2016-17; 114613/2018-4), and University Research and Scientific Production Support Program of the Goias State University (PROBIP/UEG). Phillip T. Soares provided essential help assembling the trait database. Prof. Dr. Bruno Vilela de Moraes e Silva (UFBA) read critically the first version of this manuscript and provided insightful comments that improved it.

References

Amado TF, Bidau CJ, Olalla-Tárraga MÁ (2019) Geographic variation of body size in New World anurans: energy and water in a balance. Ecography 42(3):456–466. https://doi.org/10.1111/ecog.03889

Antonelli A, Nylander JA, Persson C, Sanmartin I (2009) Tracing the impact of the Andean uplift on Neotropical plant evolution. Proc Natl Acad Sci U S A 106(24):9749–9754. https://doi.org/10.1073/pnas.0811421106

Antonelli A, Zizka A, Carvalho FA, Scharn R, Bacon CD, Silvestro D, Condamine FL (2018a) Amazonia is the primary source of Neotropical biodiversity. Proc Natl Acad Sci U S A 115(23):6034–6039. https://doi.org/10.1073/pnas.1713819115

Antonelli A, Kissling WD, Flantua SGA, Bermúdez MA, Mulch A, Muellner-Riehl AN, Kreft H, Linder HP, Badgley C, Fjeldså J, Fritz SA, Rahbek C, Herman F, Hooghiemstra H, Hoorn C (2018b) Geological and climatic influences on mountain biodiversity. Nat Geosci 11(10):718–725. https://doi.org/10.1038/s41561-018-0236-z

Badgley C, Smiley TM, Terry R, Davis EB, DeSantis LR, Fox DL, Hopkins SS, Jezkova T, Matocq MD, Matzke N, McGuire JL, Mulch A, Riddle BR, Roth VL, Samuels JX, Stromberg CA, Yanites BJ (2017) Biodiversity and topographic complexity: modern and geohistorical perspectives. Trends Ecol Evol 32(3):211–226. https://doi.org/10.1016/j.tree.2016.12.010

Both C, Cechin SZ, Melo AS, Hartz SM (2011) What controls tadpole richness and guild composition in ponds in subtropical grasslands? Austral Ecol 36(5):530–536. https://doi.org/10.1111/j.1442-9993.2010.02183.x

Chase JM, Leibold MA (2003) Ecological Niches, linking classical and contemporary approaches. Interspecific interactions. Chicago University Press, Chicago

Cornelissen J, Lavorel S, Garnier E, Diaz S, Buchmann N, Gurvich D, Reich P, Ter Steege H, Morgan H, Van Der Heijden M (2003) A handbook of protocols for standardised and easy measurement of plant functional traits worldwide. Aust J Bot 51(4):335–380

Da Silva FR, Almeida-Neto M, do Prado VHM, Haddad CFB, de Cerqueira Rossa-Feres D (2012) Humidity levels drive reproductive modes and phylogenetic diversity of amphibians in the Brazilian Atlantic Forest. J Biogeogr 39(9):1720–1732. https://doi.org/10.1111/j.1365-2699.2012.02726.x

Ernst R, Linsenmair KE, Rödel M-O (2006) Diversity erosion beyond the species level: dramatic loss of functional diversity after selective logging in two tropical amphibian communities. Biol Conserv 133(2):143–155. https://doi.org/10.1016/j.biocon.2006.05.028

Ernst R, Linsenmair KE, Thomas R, Rödel M-O (2007) Amphibian communities in disturbed forests: lessons from the Neo- and Afrotropics. In: Tscharntke T, Leuschner C, Zeller M, Guhardja E, Bidin A (eds) The stability of tropical rainforest margins, linking ecological, economic and social constraints of land use and conservation, Environmental Science and Engineering. Springer, Berlin, pp 59–85. https://doi.org/10.1007/978-3-540-30290-2_4

Ernst R, Keller A, Landburg G, Grafe TU, Linsenmair KE, Rödel M-O, Dziock F (2012) Common ancestry or environmental trait filters: cross-continental comparisons of trait-habitat relationships in tropical anuran amphibian assemblages. Glob Ecol Biogeogr 21(7):704–715. https://doi.org/10.1111/j.1466-8238.2011.00719.x

Froese R, Pauly D (2000) FishBase 2000: concepts, design and data sources. ICLARM, Los Baños/Laguna/Philippines

Gallmetzer N, Schulze CH (2015) Impact of oil palm agriculture on understory amphibians and reptiles: a Mesoamerican perspective. Global Ecology and Conservation 4:95–109. https://doi.org/10.1016/j.gecco.2015.05.008

Garnier E, Cortez J, Billès G, Navas M-L, Roumet C, Debussche M, Laurent G, Blanchard A, Aubry D, Bellmann A, Neill C, Toussaint J-P (2004) Plant functional markers capture ecosystem properties during secondary succession. Ecology 85(9):2630–2637. https://doi.org/10.1890/03-0799

Gomez-Mestre I, Pyron RA, Wiens JJ (2012) Phylogenetic analyses reveal unexpected patterns in the evolution of reproductive modes in frogs. Evolution 66(12):3687–3700. https://doi.org/10.1111/j.1558-5646.2012.01715.x

Gouveia SF, Bovo RP, Rubalcaba JG, Da Silva FR, Maciel NM, Andrade DV, Martinez PA (2019) Biophysical modeling of water economy can explain geographic gradient of body size in anurans. Am Nat 193(1):51–58. https://doi.org/10.1086/700833

Graham CH, Carnaval AC, Cadena CD, Zamudio KR, Roberts TE, Parra JL, McCain CM, Bowie RCK, Moritz C, Baines SB, Schneider CJ, VanDerWal J, Rahbek C, Kozak KH, Sanders NJ (2014) The origin and maintenance of montane diversity: integrating evolutionary and ecological processes. Ecography 37(8):711–719. https://doi.org/10.1111/ecog.00578

Grenié M, Denelle P, Tucker CM, Munoz F, Violle C, Merow C (2017) Funrar: an R package to characterize functional rarity. Divers Distrib 23(12):1365–1371. https://doi.org/10.1111/ddi.12629

Guedes TB, Azevedo JAR, Bacon C, Provete DB, Antonelli A (2019) Diversity, endemism, and evolutionary history of montane biotas outside the Andean region. In: Rull V, Carnaval A (eds) Neotropical diversification. Springer, Berlin

Hoorn C, Wesselingh FP, Steege H, Bermudez MA, Mora A, Sevink J, Sanmartín I, Sanchez-Meseguer A, Anderson CL, Figueiredo JP, Jaramillo C, Riff D, Negri FR, Hooghiemstra H, Lundberg J, Stadler T, Särkinen T, Antonelli A (2010) Amazonia through time: Andean uplift, climate change, landscape evolution, and biodiversity. Science 330:927–931

Hortal J, Bello FD, Diniz-Filho JAF, Lewinsohn TM, Lobo JM, Ladle RJ (2015) Seven shortfalls that beset large-scale knowledge on biodiversity. Annual Review of Ecology, Evolution, and Systematics 46(1):523–549. https://doi.org/10.1146/annurev-ecolsys-112414-054400

Hutter CR, Guayasamin JM, Wiens JJ (2013) Explaining Andean megadiversity: the evolutionary and ecological causes of glassfrog elevational richness patterns. Ecol Lett 16(9):1135–1144. https://doi.org/10.1111/ele.12148

Hutter CR, Lambert SM, Wiens JJ (2017) Rapid diversification and time explain amphibian richness at different scales in the Tropical Andes, Earth's most biodiverse hotspot. Am Nat 190(6):828–843. https://doi.org/10.1086/694319

Jones KE, Bielby J, Cardillo M, Fritz SA, O'Dell J, Orme CDL, Safi K, Sechrest W, Boakes EH, Carbone C, Connolly C, Cutts MJ, Foster JK, Grenyer R, Habib M, Plaster CA, Price SA, Rigby EA, Rist J, Teacher A, Bininda-Emonds ORP, Gittleman JL, Mace GM, Purvis A, Michener WK (2009) PanTHERIA: a species-level database of life history, ecology, and geography of extant and recently extinct mammals. Ecology 90(9):2648–2648. https://doi. org/10.1890/08-1494.1

Kruk C, Huszar VL, Peeters ET, Bonilla S, Costa L, Lürling M, Reynolds CS, Scheffer M (2010) A morphological classification capturing functional variation in phytoplankton. Freshw Biol 55(3):614–627

Kuebbing SE, Maynard DS, Bradford MA, Huenneke L (2017) Linking functional diversity and ecosystem processes: a framework for using functional diversity metrics to predict the ecosystem impact of functionally unique species. J Ecol 106:687. https://doi. org/10.1111/1365-2745.12835

Laliberté E, Legendre P (2010) A distance-based framework for measuring functional diversity from multiple traits. Ecology 91(1):299–305

Laliberté E, Legendre P, Shipley B (2014) FD: measuring functional diversity from multiple traits, and other tools for functional ecology. R package version 1.0-12

Lefcheck JS, Bastazini VAG, Griffin JN (2014) Choosing and using multiple traits in functional diversity research. Environ Conserv 42(02):104–107. https://doi.org/10.1017/ s0376892914000307

Legendre P, Legendre L (2012) Numerical ecology, 3rd edn. Elsevier Limited, Oxford

Legras G, Gaertner J-C (2018) Assessing functional evenness with the FEve index: a word of warning. Ecol Indic 90:257–260. https://doi.org/10.1016/j.ecolind.2018.03.020

Legras G, Loiseau N, Gaertner JC (2018) Functional richness: overview of indices and underlying concepts. Acta Oecol 87:34–44. https://doi.org/10.1016/j.actao.2018.02.007

Lescano JN, Miloch D, Leynaud GC (2018) Functional traits reveal environmental constraints on amphibian community assembly in a subtropical dry forest. Austral Ecol 43:623. https://doi. org/10.1111/aec.12607

Lion MB, Mazzochini GG, Garda AA, Lee TM, Bickford D, Costa GC, Fonseca CR, Algar A (2019) Global patterns of terrestriality in amphibian reproduction. Glob Ecol Biogeogr 28:744. https://doi.org/10.1111/geb.12886

Lipinski VM, Iop S, Schuch AP, Santos TG (2018) Enhanced phylogenetic diversity of anuran communities: a result of species loss in an agricultural environment. In: Sudarshana P (ed) Tropical forests – new edition. Intech open, London

Litchman E, Klausmeier CA (2008) Trait-based community ecology of phytoplankton. Annu Rev Ecol Evol Syst 39(1):615–639. https://doi.org/10.1146/annurev.ecolsys.39.110707.173549

Litchman E, Ohman MD, Kiørboe T (2013) Trait-based approaches to zooplankton communities. J Plankton Res 35(3):473–484

Lomolino MV, Riddle BR, Wittaker RJ (2016) Biogeography, 5th edn. Sinauer, Sunderland, MA

Májeková M, Paal T, Plowman NS, Bryndova M, Kasari L, Norberg A, Weiss M, Bishop TR, Luke SH, Sam K, Le Bagousse-Pinguet Y, Leps J, Gotzenberger L, de Bello F (2016) Evaluating functional diversity: missing trait data and the importance of species abundance structure and data transformation. PLoS One 11(2):e0149270. https://doi.org/10.1371/journal.pone.0149270

Mason NWH, de Bello F (2013) Functional diversity: a tool for answering challenging ecological questions. J Veg Sci 24(5):777–780. https://doi.org/10.1111/Jvs.12097

Mason NWH, Mouillot D, Lee WG, Wilson JB (2005) Functional richness, functional evenness and functional divergence: the primary components of functional diversity. Oikos 111(1):112–118. https://doi.org/10.1111/j.0030-1299.2005.13886.x

Mason NWH, de Bello F, Mouillot D, Pavoine S, Dray S (2013) A guide for using functional diversity indices to reveal changes in assembly processes along ecological gradients. J Veg Sci 24(5):794–806. https://doi.org/10.1111/jvs.12013

McGill BJ, Enquist BJ, Weiher E, Westoby M (2006) Rebuilding community ecology from functional traits. Trends in Ecology and Evolution 21(4):178–185

McPherson JM, Yeager LA, Baum JK, Price S (2018) A simulation tool to scrutinise the behaviour of functional diversity metrics. Methods Ecol Evol 9(1):200–206. https://doi. org/10.1111/2041-210x.12855

Merckx VS, Hendriks KP, Beentjes KK, Mennes CB, Becking LE, Peijnenburg KT, Afendy A, Arumugam N, de Boer H, Biun A, Buang MM, Chen PP, Chung AY, Dow R, Feijen FA, Feijen H, Feijen-van Soest C, Geml J, Geurts R, Gravendeel B, Hovenkamp P, Imbun P, Ipor I, Janssens SB, Jocque M, Kappes H, Khoo E, Koomen P, Lens F, Majapun RJ, Morgado LN, Neupane S, Nieser N, Pereira JT, Rahman H, Sabran S, Sawang A, Schwallier RM, Shim PS, Smit H, Sol N, Spait M, Stech M, Stokvis F, Sugau JB, Suleiman M, Sumail S, Thomas DC, van Tol J, Tuh FY, Yahya BE, Nais J, Repin R, Lakim M, Schilthuizen M (2015) Evolution of endemism on a young tropical mountain. Nature 524(7565):347–350. https://doi.org/10.1038/nature14949

Mittermeier RA, Gil PR, Hoffmann M, Pilgrim J, Brooks T, Mittermeier CG, Lamourex J, Fonseca GAB (2005) Hotspots revisited. Earth's biologically richest and most endangered terrestrial ecorregions. CEMEX, Ciudad de México

Moretti M, Dias ATC, de Bello F, Altermatt F, Chown SL, Azcárate FM, Bell JR, Fournier B, Hedde M, Hortal J, Ibanez S, Öckinger E, Sousa JP, Ellers J, Berg MP, Fox C (2017) Handbook of protocols for standardized measurement of terrestrial invertebrate functional traits. Funct Ecol 31(3):558–567. https://doi.org/10.1111/1365-2435.12776

Müller H, Liedtke HC, Menegon M, Beck J, Ballesteros-Mejia L, Nagel P, Loader SP (2013) Forests as promoters of terrestrial life-history strategies in East African amphibians. Biol Lett 9(3):20121146. https://doi.org/10.1098/rsbl.2012.1146

Myhrvold NP, Baldridge E, Chan B, Sivam D, Freeman DL, Ernest SKM (2015) An amniote life-history database to perform comparative analyses with birds, mammals, and reptiles. Ecology 96(11):3109–3000. https://doi.org/10.1890/15-0846r.1

Newbold T, Butchart SH, Sekercioglu CH, Purves DW, Scharlemann JP (2012) Mapping functional traits: comparing abundance and presence-absence estimates at large spatial scales. PLoS One 7(8):e44019. https://doi.org/10.1371/journal.pone.0044019

Nowakowski AJ, Thompson ME, Donnelly MA, Todd BD (2017) Amphibian sensitivity to habitat modification is associated with population trends and species traits. Glob Ecol Biogeogr 26(6):700–712. https://doi.org/10.1111/geb.12571

Oliveira BF, Scheffers BR (2018) Vertical stratification influences global patterns of biodiversity. Ecography 42(2):249–249. https://doi.org/10.1111/ecog.03636

Oliveira BF, Machac A, Costa GC, Brooks TM, Davidson AD, Rondinini C, Graham CH (2016) Species and functional diversity accumulate differently in mammals. Glob Ecol Biogeogr 25:1119. https://doi.org/10.1111/geb.12471

Oliveira BF, Sao-Pedro VA, Santos-Barrera G, Penone C, Costa GC (2017) AmphiBIO, a global database for amphibian ecological traits. Sci Data 4:170123. https://doi.org/10.1038/sdata.2017.123

Olson DM, Dinerstein E, Wikramanayake ED, Burgess ND, Powell GVN, Underwood EC, D'amico JA, Itoua I, Strand HE, Morrison JC, Loucks CJ, Allnutt TF, Ricketts TH, Kura Y, Lamoreux JF, Wettengel WW, Hedao P, Kassem KR (2001) Terrestrial ecoregions of the world: a new map of life on earth. Bioscience 51(11):933–938

Ouchi-Melo LS, Meynard CN, Gonçalves-Souza T, de Cerqueira R-FD (2018) Integrating phylogenetic and functional biodiversity facets to guide conservation: a case study using anurans in a global biodiversity hotspot. Biodivers Conserv 27(12):3247–3266

Pavoine S, Vallet J, Dufour A-B, Gachet S, Daniel H (2009) On the challenge of treating various types of variables: application for improving the measurement of functional diversity. Oikos 118(3):391–402. https://doi.org/10.1111/j.1600-0706.2008.16668.x

Pebesma E (2018) Simple features for R: standardized support for spatial vector data. The R Journal. https://journal.r-project.org/archive/2018/RJ-2018-009/

Pérez-Harguindeguy N, Diaz S, Gamier E, Lavorel S, Poorter H, Jaureguiberry P, Bret-Harte M, Comwell W, Craine J, Gurvich D (2013) New handbook for standardised measurement of plant functional traits worldwide. Aust J Bot 61:167–234

Peters RH (1986) The ecological implications of body size. Cambridge University Press, Cambridge

Queiroz CS, da Silva FR, Rossa-Feres DC (2015) The relationship between pond habitat depth and functional tadpole diversity in an agricultural landscape. R Soc Open Sci 2(7):150165. https://doi.org/10.1098/rsos.150165

Rangel TF, Edwards NR, Holden PB, Diniz-Filho JAF, Gosling WD, Coelho MTP, Cassemiro FAS, Rahbek C, Colwell RK (2018) Modeling the ecology and evolution of biodiversity: biogeographical cradles, museums, and graves. Science 361(6399):eaar5452. https://doi.org/10.1126/science.aar5452

Reynolds CS, Huszar V, Kruk C, Naselli-Flores L, Melo S (2002) Towards a functional classification of the freshwater phytoplankton. J Plankton Res 24(5):417–428. https://doi.org/10.1093/plankt/24.5.417

Riemann JC, Ndriantsoa Serge H, Rödel M-O, Glos J (2017) Functional diversity in a fragmented landscape — Habitat alterations affect functional trait composition of frog assemblages in Madagascar. Global Ecology and Conservation 10:173–183. https://doi.org/10.1016/j.gecco.2017.03.005

Rosado BHP, Figueiredo MSL, de Mattos EA, Grelle CEV (2016) Eltonian shortfall due to the Grinnellian view: functional ecology between the mismatch of niche concepts. Ecography 39(11):1034–1041. https://doi.org/10.1111/ecog.01678

Rubalcaba JG, Gouveia SF, Olalla-Tárraga MA, Algar A (2019) A mechanistic model to scale up biophysical processes into geographical size gradients in ectotherms. Glob Ecol Biogeogr 28:793. https://doi.org/10.1111/geb.12893

Schleuter D, Daufresne M, Massol F, Argillier C (2010) A user's guide to functional diversity indices. Ecol Monogr 80(3):469–484. https://doi.org/10.1890/08-2225.1

Sepulchre P, Sloan LC, Fluteau F (2011) Modelling the response of Amazonian climate to the uplift of the Andean mountain range. In: Hoorn C, Wesselingh F (eds) Amazonia: landscape and species evolution: a look into the past. Wiley, New York

Slavenko A, Meiri S (2015) Mean body sizes of amphibian species are poorly predicted by climate. J Biogeogr 42(7):1246–1254. https://doi.org/10.1111/jbi.12516

Strauß A, Reeve E, Randrianiaina R-D, Vences M, Glos J (2010) The world's richest tadpole communities show functional redundancy and low functional diversity: ecological data on Madagascar's stream-dwelling amphibian larvae. BMC Ecol 10(1):12. https://doi.org/10.1186/1472-6785-10-12

Strauß A, Guilhaumon F, Randrianiaina RD, Wollenberg Valero KC, Vences M, Glos J (2016) Opposing patterns of seasonal change in functional and phylogenetic diversity of tadpole assemblages. PLoS One 11(3):e0151744. https://doi.org/10.1371/journal.pone.0151744

Trakimas G, Whittaker RJ, Borregaard MK (2016) Do biological traits drive geographical patterns in European amphibians? Glob Ecol Biogeogr 25:1228. https://doi.org/10.1111/geb.12479

Trimble MJ, van Aarde RJ (2014) Amphibian and reptile communities and functional groups over a land-use gradient in a coastal tropical forest landscape of high richness and endemicity. Anim Conserv 17(5):441–453. https://doi.org/10.1111/acv.12111

Trochet A, Moulherat S, Calvez O, Stevens V, Clobert J, Schmeller D (2014) A database of life-history traits of European amphibians. Biodivers Data J 2:e4123

Tsianou MA, Kallimanis AS (2015) Different species traits produce diverse spatial functional diversity patterns of amphibians. Biodivers Conserv 25(1):117–132. https://doi.org/10.1007/s10531-015-1038-x

Tsianou MA, Kallimanis AS (2018) Trait selection matters! A study on European amphibian functional diversity patterns. Ecological Research 34:1–10. https://doi.org/10.1111/1440-1703.1002

Vasconcelos TS, Prado VHM, da Silva FR, Haddad CFB (2014) Biogeographic distribution patterns and their correlates in the diverse frog fauna of the Atlantic Forest hotspot. PLoS One 9(8):e104130. https://doi.org/10.1371/journal.pone.0104130

Villalobos F, Dobrovolski R, Provete DB, Gouveia SF (2013) Is rich and rare the common share? Describing biodiversity patterns to inform conservation practices for South American anurans. PLoS One 8(2):e56073. https://doi.org/10.1371/journal.pone.0056073

Villéger S, Mason NW, Mouillot D (2008) New multidimensional functional diversity indices for a multifaceted framework in functional ecology. Ecology 89(8):2290–2301

Violle C, Navas M-L, Vile D, Kazakou E, Fortunel C, Hummel I, Garnier E (2007) Let the concept of trait be functional! Oikos 116(5):882–892. https://doi.org/10.1111/j.2007.0030-1299.15559.x

Violle C, Reich PB, Pacala SW, Enquist BJ, Kattge J (2014) The emergence and promise of functional biogeography. Proc Natl Acad Sci U S A 111(38):13690–13696. https://doi.org/10.1073/pnas.1415442111

Violle C, Borgy B, Choler P (2015) Trait databases: misuses and precautions. J Veg Sci 26:826–827. https://doi.org/10.1111/jvs.12325

Violle C, Thuiller W, Mouquet N, Munoz F, Kraft NJ, Cadotte MW, Livingstone SW, Mouillot D (2017) Functional rarity: the ecology of outliers. Trends in Ecology and Evolution 32(5):356–367

Vonhof HB, Kaandorp RJG (2011) Climate variation in Amazonia during the Neogene and the quaternary. In: Hoorn C, Wesselingh F (eds) Amazonia: landscape and species evolution: a look into the past. Wiley, New York

Wickham H (2016) ggplot2: elegant graphics for data analysis, 2nd edn. Springer, New York

Wiens JJ, Pyron RA, Moen DS (2011) Phylogenetic origins of local-scale diversity patterns and the causes of Amazonian megadiversity. Ecol Lett 14(7):643–652. https://doi.org/10.1111/j.1461-0248.2011.01625.x

Wilman H, Belmaker J, Simpson J, de la Rosa C, Rivadeneira MM, Jetz W (2014) EltonTraits 1.0: species-level foraging attributes of the world's birds and mammals. Ecology 95(7):2027–2027

Zhu L, Fu B, Zhu H, Wang C, Jiao L, Zhou J (2017) Trait choice profoundly affected the ecological conclusions drawn from functional diversity measures. Sci Rep 7(1):3643. https://doi.org/10.1038/s41598-017-03812-8

Zuloaga J, Currie DJ, Kerr JT, Pither J (2019) The origins and maintenance of global species endemism. Glob Ecol Biogeogr 28(2):170–183. https://doi.org/10.1111/geb.12834

Chapter 6
Biogeographic Regionalization of South American Anurans

Abstract The interest in recognizing spatial patterns of species co-distributions has long led biogeographers and macroecologists to classify the world in biogeographic regions. In this chapter, we aimed to identify regions with distinct species pools, thus representing different biogeographic regions with co-occurring species of anurans in South America. Using quantitative and clustering methods, we recognized six anuran biogeographic regions in South America: two regions are predominantly tropical (named as AMAZON and DIAGONAL-AF); two regions are associated to the Andes mountains (named as MID-ANDES and NORTH-/SOUTH-ANDES); and two regions are broadly located south of the Tropic of Capricorn (named as SUB-TROPICAL and TEMP-GRASS). Using regression and variation partitioning analyses, the six distinct biogeographic regions are mainly predicted by differences in climatic gradients among the biogeographic regions (e.g., clusters located in the different tropical, subtropical, and temperate regions). Yet, the combination of rough topography and habitat structure of major biomes was also a good predictor for other biogeographic regions (e.g., the recognition of the different Andean biogeographic regions having different major biomes, such as montane forests and grasslands).

Keywords Anura · Biogeographic regions · Bioregionalization · Climate hypothesis · recluster.region · South America

6.1 Introduction

Biological diversity is not distributed randomly on Earth. The geographic distribution of species is constrained by different evolutionary and ecological processes in a hierarchical manner. For instance, organisms may occupy a new location through dispersal events and, once established, may evolve as new species through different speciation processes. Then, different environmental and ecological constraints modulate their specific occurrences, thus adjusting the species' geographical ranges

© Springer Nature Switzerland AG 2019
T. S. Vasconcelos et al., *Biogeographic Patterns of South American Anurans*,
https://doi.org/10.1007/978-3-030-26296-9_6

and generating different patterns of distribution (Booth and Swanton 2002; Lomolino et al. 2017). The outcome is that a given region will support co-occurring species having either similar or different geographic distribution patterns among one another. Then, uncovering the mechanisms that led species to share similar geographic patterns has been an interesting subject for biogeographers to explore.

The interest in recognizing spatial patterns of species co-distributions has long led biogeographers and macroecologists to classify the world in biogeographic regions. These regions are delimited by discontinuities in the limits of the geographic distribution of species (Ferro and Morrone 2014). The periphery of these regions represents areas with well-delimited changes in current or past climatic regimes, topographic relief, structure of major habitats, or other geographic barriers (e.g., habitat disturbance, riverine barriers; Rueda et al. 2010; Ferro and Morrone 2014; Vasconcelos et al. 2014; Ficetola et al. 2017; Godinho and da Silva 2018). Then, the identification of biogeographic regions is the first step to uncover the mechanisms that may have driven current species distributions.

South America is the most biological diverse continent on Earth (Rangel et al. 2018 and references therein). As mentioned in Chap. 1, South America originated from the western Gondwanan landmass that separated from Pangea throughout the Jurassic to mid-Cretaceous (~200–100 Mya) and has undergone several geologic and climate change events ever since (see review in Lomolino et al. 2017). These events, along with recent climate conditions, have been attributed as the main drivers of the current geographic ranges of South American species (Antonelli et al. 2018; Rangel et al. 2018). Nonetheless, disentangling the relative contribution of such variables in establishing the geographical pattern of a specific taxon or group of species has remained a challenge to biogeographers.

As pointed in previous chapters, anuran amphibians are a highly diverse vertebrate group in South America with increasing species description rates along the years (see Chap. 2). The different facets of anuran diversity explored in the previous chapters of this book have recognized different regions in the continent as having high diversity values associated to tropical forested and/or high-altitude regions, whereas low diversity areas are mainly concentrated in subtropical and temperate regions, with tropical open formations having intermediate diversity values.

In this chapter, we verified whether regions with the highest, lowest, and intermediate values of diversity are composed of distinct species pools, thus representing different biogeographic regions with co-occurring species. Then, we propose a regionalization scheme for the South American anuran fauna using quantitative and clustering methods to identify co-occurrence of species distributions within the continental extent. Subsequently, we identified clusters' correlates testing for the following non-exclusive hypotheses:

1. Climate hypothesis – anurans are highly dependent on climate conditions due to their complex life cycle that generally involves an aquatic larvae and a terrestrial adult phase that broadly depends on the humid atmosphere to maintain effective gas exchange throughout their skins (Duellman and Trueb 1994; Wells 2007). Thus, biogeographic regions correlated with current climatic variables may indicate species pools adapted to different climate regimes.

2. Topography hypothesis – rough topographies are assumed to restrict species ranges and generate species diversification through allopatric speciation for a range of organisms (Ruggiero and Hawkins 2008; Antonelli et al. 2009; Rangel et al. 2018). Thus, biogeographic regions correlated with topographic variables may indicate co-occurring species having their distributional ranges strongly constrained by topographic barriers.
3. Vegetation structure hypothesis – the habitat provides the templet on which evolution forges animal life-history strategies (Southwood 1977; Rueda et al. 2010). For instance, the high endemism rates of Atlantic Forest anurans (80% of endemism) are attributed to the humid microhabitat provided by the ombrophilous forests, which in turn promoted the adaptation of a diverse number of reproductive modes (da Silva et al. 2012; Haddad et al. 2013). Thus, biogeographic regions correlated with major biomes in South America may indicate species pools adapted to the respective habitat structure.

6.2   Material and Methods

The presence/absence matrix of anurans in South America, detailed in Chap. 1, was submitted to the *recluster.region* algorithm. This algorithm was chosen among a variety of options (see review in Kreft and Jetz 2010) because it offers interesting tools, as described ahead, to identify the optimal number of clusters (e.g., Holt et al. 2013; Godinho and da Silva 2018). The *recluster.region* algorithm (Dapporto et al. 2013, 2015) initially calculates species composition dissimilarities between pairs of grid cells. Since the variation in species composition within South America is broadly composed of the species turnover among the grid cells, we chose the Simpson index that represents a better fit to the data with high species replacement among grids (Kreft and Jetz 2010). We produced 50 trees (triangular dissimilarity matrices), so this means that the grids were randomly reordered from the original matrix to minimize the effect that the order of areas in the original presence-absence matrix has on the final topology of the dissimilarity matrix. This procedure was then adopted because the high frequency of ties and zeros produced by beta-diversity indices affects the topology and bootstrap support of dendrograms generated from the dissimilarity matrix (Dapporto et al. 2013). The algorithm also identifies clusters according to a predefined number, so we set a minimum of 2 and a maximum of 50 clusters. We selected the optimal number of clusters based on the explained dissimilarity and the mean silhouette width (Holt et al. 2013; Borcard et al. 2011; Godinho and da Silva 2018). In summary, the explained dissimilarity values show the maximization of the between-cluster variation relative to the within-cluster variation, so solutions with *n* clusters whose explained dissimilarity threshold values reach ~90% are an appropriate choice for establishing a suitable cut (Holt et al. 2013; Godinho and da Silva 2018). Conversely, the mean silhouette width measures the robustness, i.e., whether grid cells are located in the correct cluster by measuring the strengths of any partition of objects from a dissimilarity matrix (Borcard et al.

Table 6.1 Cumulative eigenvalues of the five axis of the principal component analysis of the climatic variables considered

	Axis 1	Axis 2	Axis 3	Axis 4	Axis 5
AT	0.4813	0.4825	0.6591	0.9962	1.00
TS	0.9752	0.9999	0.9999	1.00	1.00
PS	0.0043	0.3957	0.4253	0.5093	1.00
AP	0.5676	0.9971	0.9999	0.9999	1.00
AET	0.4261	0.5861	0.9998	0.9999	1.00

AT average annual temperature, *TS* temperature seasonality, *PS* precipitation seasonality, *AP* annual precipitation, *AET* actual evapotranspiration

2011; Godinho and da Silva 2018). Finally, we first identified the number of clusters that reached the threshold value of 90%, and, then, we selected the cluster number when the mean silhouette value stopped increasing, following Holt et al. (2013) and Godinho and da Silva (2018).

The climatic variables used to identifying the correlates of biogeographical regions were annual precipitation, precipitation seasonality, average annual temperature, temperature seasonality, and actual evapotranspiration (see details in Chap. 1). To reduce the number of variables and to avoid collinearity, we performed a principal component analysis (PCA) and considered the first two axes (that contained 98.7% of the variation, Table 6.1) as a representation of the selected climatic variables. Further, we tested a series of multinomial logistic regression models to identify the correlates of biogeographic regions. Models were built combining the two first PCA axes of climatic variables, the topographic variable, and the major habitat structure variable, all of them described in Chap. 1. Specifically, we combined single and multiple variables in models that could potentially explain the clusters' geographic pattern (see similar approach in Vasconcelos et al. 2014 and Godinho and da Silva 2018). The best-fit model was selected by the lowest Akaike's Information Criterion corrected for small sample size (AICc; Burnham and Anderson 2002). Then, we submitted the best-fit model to the variation partitioning analysis (Borcard et al. 1992) in order to verify the independent and shared effects of climate, topography, and major biome structure on the biogeographical regionalization scheme.

6.3 Results and Discussion

We found that the optimal number of biogeographic regions for South American anurans was six (Fig. 6.1, Table 6.2). Two regions are predominantly tropical: (a) a region that broadly encompasses the Amazon domain, hereafter identified as AMAZON, and (b) a region that broadly encompasses the northeast-southwest "dry diagonal" of open formations, namely, Caatinga-Cerrado-Chaco, along with its eastward forest formations of the Atlantic Forest domain, hereafter identified as DIAGONAL-AF. Two regions are either tropical or subtropical/temperate: (a) a high-altitude region encompassing a contiguous area within the central Andes,

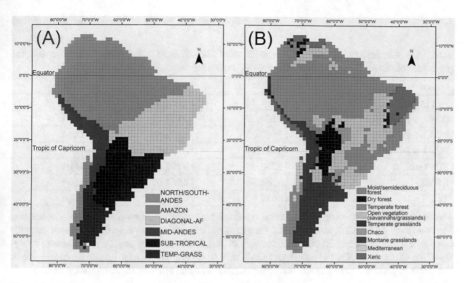

Fig. 6.1 (a) Biogeographic regions based on the South American anuran fauna submitted to the recluster.region algorithm (see Methods); (b) Major South American biomes modified from the World Wildlife Fund designations (Olson et al. 2001)

Table 6.2 Values of the mean silhouette width (SW) and the explained dissimilarity (ED) for all cluster solutions

Number of cluster	SW	ED	Number of cluster	SW	ED
2	0.2540	0.5426	27	0.3144	0.9881
3	0.3252	0.7558	28	0.3175	0.9886
4	0.3299	0.8235	29	0.3227	0.9893
5	0.3566	0.8471	30	0.3203	0.9895
6	**0.3718**	**0.8798**	31	0.3241	0.9900
7	0.3657	0.8891	32	0.3335	0.9904
8	0.3520	0.8979	33	0.3347	0.9904
9	0.2739	0.9255	34	0.3344	0.9909
10	0.2602	0.9333	35	0.2945	0.9910
11	0.2759	0.9481	36	0.3006	0.9914
12	0.2878	0.9522	37	0.2949	0.9919
13	0.2916	0.9553	38	0.3154	0.9922
14	0.2929	0.9659	39	0.3266	0.9928
15	0.3155	0.9698	40	0.3298	0.9930
16	0.3132	0.9725	41	0.3240	0.9933
17	0.2924	0.9765	42	0.3158	0.9935
18	0.2860	0.9777	43	0.3164	0.9937
19	0.3014	0.9798	44	0.3149	0.9940
20	0.3094	0.9811	45	0.3192	0.9942
21	0.3111	0.9830	46	0.3202	0.9943
22	0.3151	0.9840	47	0.2644	0.9943
23	0.3231	0.9848	48	0.2703	0.9945
24	0.3191	0.9859	49	0.2741	0.9946
25	0.3209	0.9866	50	0.2924	0.9948
26	0.3119	0.9873			

In bold the clustering solutions are selected

Table 6.3 Generalized linear models with multinomial logit-link considered to assess the predictors of the six biogeographic regions found for anurans in South America

Model	AIC	AICc	wAICc
Full model	1917.74	1918.54	1.0000
PCA1 + PCA2 + TOP	2104.19	2104.71	<0.001
PCA1 + PCA2 + BIOM	2562.06	2562.58	<0.001
PCA1 + PCA2	2707.11	2707.41	<0.001
TOP+BIOM	3798.38	3798.67	<0.001
BIOM	4566.94	4567.07	<0.001
TOP	4618.90	4619.04	<0.001

Models are sorted according to the lowest Akaike information criterion (AIC) and AIC corrected for small samples (AICc). wAICc, AICc weight model expressing the weight of evidence supporting the best model among all others compared. PCA1, first axis of the principal component analysis of the climatic variables considered; PCA2, second axis of the principal component analysis of the climatic variables considered; TOP, topographic variable; BIOM, major habitat structure variable

Fig. 6.2 Deviance partitioning analysis representing the deviance in the clusters configuration explained by climate, topography, and major biomes of South America

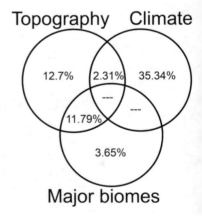

hereafter identified as MID-ANDES, and (b) a discontinuous region in the north and southern Andes, hereafter identified as NORTH/SOUTH-ANDES. The other two regions are broadly located south of the Tropic of Capricorn: (a) a subtropical region composed of both open (Pampa biome and part of the temperate grasslands biome) and forest formations (south of the Atlantic Forest domain and part of the Chaco/dry forest biome), hereafter identified as SUB-TROPICAL, and (b) a region mostly in the southern temperate grasslands, hereafter identified as TEMP-GRASS.

The best-fit model for the six-cluster solution was the model that combined all variables (weigh evidence of 1.00; Table 6.3). The strongest correlate of the clusters' geographic pattern was climate (35.34%), followed by topography (12.17%), and habitat structure of major biomes (3.65%; Fig. 6.2). Although the independent effect of major biomes explained a small fraction, this variable has a high shared effect with topography (11.79%), whereas topography has also a shared effect with climate that accounts for 2.31% of the variability in the cluster pattern (Fig. 6.2).

Taken together, these results suggest that the six distinct biogeographic regions are mainly predicted by differences in climatic gradients among the biogeographic regions (e.g., clusters located in the different tropical, subtropical, and temperate regions). Yet, combinations of rough topography and habitat structure of major biomes were also good predictors for other biogeographic regions (e.g., the recognition of the different Andean biogeographic regions having different major biomes, such as montane forests and grasslands).

Using a wide range of taxa as biological models (e.g., insects, amphibians, birds, and mammals), previous biogeographical schemes for South America generally split the continent into two or three regions, in which there is the clear identification of the Neotropical region (sometimes called Guianan-Brazilian subregion, but see Morrone 2014) and the Andean-Patagonian region (also known as Austral, Argentinean, and Chilean, among others; Proches 2005; Kreft and Jetz 2010; Rueda et al. 2013; Morrone 2015, 2017, and references therein). Here, we found the Neotropical region within a combination of tropical clusters (AMAZON and DIAGON-AF) and the Andean region as a combination of the tropical/subtropical (MID-ANDES) and temperate clusters (TEMP-GRASS). Therefore, our regionalization proposal with more regions than the typical two or three from previous studies may be due to two non-exclusive factors: (a) the different geographic extent among the studies and (b) the methodological tools to estimate biogeographic regions. Considering the first factor, global analyses using similar clustering methods have recurrently identified Europe as part of the Palearctic region (Kreft and Jetz 2010; Rueda et al. 2013). However, scaling down the geographic extent to the continent identifies many subregions (Rueda et al. 2010). Similarly, previous proposals for South American regionalization that typically identified two or three biogeographic regions (e.g., Proches 2005; Kreft and Jetz 2010; Rueda et al. 2013; Vilhena and Antonelli 2015; Morrone 2017, and references therein) were basically performed at the global scale.

The methodological issue can be exemplified by a previous amphibian regionalization scheme performed for South America that identified three and four biogeographic regions (Vasconcelos et al. 2011). The different clustering solutions between Vasconcelos et al. (2011) and the present study may certainly be due to the use of an enhanced dataset that included newly described species and the use of different analytical tools to identify the clustering patterns. However, both Vasconcelos et al. (2011) and this study identified the AMAZON and DIAGONAL-AF as distinct anuran biogeographic units. Conversely, the single cluster identified by Vasconcelos et al. (2011) encompassing the Andes and subtropical and temperate open domains is herein split into four clusters.

Amazonia is a complex region encompassing areas with distinct evolutionary events, which in turn generated different distributional patterns for many taxa (Morrone 2017 and references therein, Godinho and da Silva 2018). Moreover, Amazonia has a historical importance in which species evolved and dispersed to other Neotropical regions, being thus considered an important source of biodiversity in South America (Antonelli et al. 2018). Besides all complexity within Amazonia, the limits of this tropical moist forest clearly show that the habitat structure has an important role in maintaining broad-scale biogeographic patterns. This is true mainly when we contrast our results with previous studies that identified the cluster AMAZON as a biogeographic unit for South American amphibians (Vasconcelos et al. 2011; Vilhena and Antonelli 2015). In addition to the habitat structure, the overall stability of humidity and moderate-to-high temperature of this tropical forest probably favored anurans to explore and breed through different forest microhabitats, which in turn led anurans to a high diversity within the AMAZON (~600 species: Lima et al. 2006; Godinho and da Silva 2018).

The cluster DIAGONAL-AF is the second pattern we found that is congruent with the amphibian regionalization scheme proposed by Vasconcelos et al. (2011) and Vilhena and Antonelli (2015). An intriguing issue here is the inclusion of distinct major habitats as a single biogeographic unit (open formations of the Caatinga-Cerrado-Chaco complex vs the central-northern Atlantic Forest). These morphoclimatic domains have distinct climate regimes, in which open formation areas have higher temperatures, less precipitation volumes, and remarkable seasonality in temperature and rainfall compared to the Atlantic Forest (IBGE 2012). These differences make anuran composition very different among those regions, with distinct adaptations to open vs forest formations within the DIAGONAL-AF. For instance, the Atlantic Forest has approximately 550 anuran species with many reproductive modes and ~80% of endemic species (Haddad et al. 2013), whereas only 209 species occur in Cerrado, of which 52% are endemic (Valdujo et al. 2012) and with considerably lower diversity of reproductive modes, probably due to harsher climatic conditions that decrease the availability of humid microhabitats for anuran reproduction (Vasconcelos et al. 2010). Yet, the Cerrado and the Atlantic Forest share many species. For example, semidecidual Atlantic Forests share a high proportion of anuran species with Cerrado, but not with the Atlantic Rain Forests, because Cerrado and semideciduous Atlantic Forests have a fairly common seasonal climate (Pennington et al. 2000; Santos et al. 2009; Vasconcelos et al. 2010).

We identified the clusters MID-ANDES and NORTH/SOUTH-ANDES as distinct anuran biogeographic units along the whole extension of the Andes. An intriguing result is that the cluster NORTH/SOUTH-ANDES is disjoint, with the northern portion – richer in anuran species – located around the equatorial zone, whereas the southern portion, with considerably lower anuran richness, is located southerly beyond of the Tropic of Capricorn. A common feature of this discontinuous cluster is that both portions have high number of micro-endemic species (see Chaps. 3 and 5 and Villalobos et al. 2013). Therefore, these small-ranged species may have influenced the recognition of these regions as disconnected units. Since most statistical tools in biogeographical regionalization have a hierarchical structure (i.e., smaller areas nested within larger ones), and that the whole Andes has many small-ranged species, the *recluster.region* algorithm will first identify areas in which some species have some congruent distribution patterns. Then, the algorithm identified the cluster MID-ANDES that have species with congruent distributions within the highland grasslands of northwestern Argentina and southern Peru. The remaining Andean areas (the cluster NORTH/SOUTH-ANDES), in which the only similarity is the scattered presence of small-ranged species, are thus identified as a new cluster (even though they are geographically disconnected). Finally, the NORTH/SOUTH-ANDES was recognized and its common share is the rarity and punctuality in the occurrence records of most species.

Specifically, the NORTH-ANDES sub-cluster is characterized by a high proportion of small-ranged species (similar to the Atlantic Forest in Southeastern Brazil, see Chapter 3, Villalobos et al. 2013) that are mostly associated with forested habitats of the Tropical Andes hotspot (Mittermeier et al. 2004). Although having small-ranged species too, the SOUTH-ANDES sub-cluster has considerable lower richness in the colder Chilean Mediterranean and temperate forests (Villalobos et al. 2013, Chapter 3). On the other hand, the MID-ANDES cluster was recognized mainly because of the predominance of highland grasslands above 3800–4500 m (Morrone 2017 and references therein), which presumably restricted the distributional ranges of some endemic species of this habitat type during the evolutionary history of South America. In summary, we recognize the Andes as a complex and heterogeneous region in which anurans have the smallest ranges in the continent and highest diversity mainly associated to high-altitude tropical forests.

The last two clusters are found southward from the Tropic of Capricorn. SUB-TROPICAL is a mix of either open vegetation or forest formations in which cold conditions and temperature seasonality become prominent environmental features. The intriguing point here is that the anuran composition in the southern Atlantic Forest (i.e., the Araucaria moist forest, sensu Olson et al. 2001; IBGE 2012) is more related to the composition of open domains of southern South America than to the core composition of the Atlantic Forest. This compositional similarity may be explained by the southern circum-Amazonia "arcs" of seasonally dry climates, sometimes patchy and sometimes continuous across the historical evolution of South America. This fact presumably formed biogeographical bridges connecting the Tropical Andes and the Atlantic Forest (Rangel et al. 2018 and references therein), so this may have resulted in a high similarity of anuran composition in this subtropical region of the continent. Lastly, the cluster TEMP-GRASS is located within the mid-southern temperate grasslands. This is the less diverse biogeographic unit in the continent (see Chaps. 3, 4, and 5), probably due to the extremely cold conditions that limited anuran occurrences to the cold-adapted species endemic to this region.

In conclusion, we recognize six biogeographic units for South American anurans which were correlated with different contemporary environmental predictors. Among these, current climatic regimes are the best predictors for the clustering pattern, followed by a combination of topography and the structure of major biomes. We emphasize that these correlates should not be viewed as direct causal links to explain the clustering pattern, but rather as a consequence of historical events driven by geographically structured climate dynamics and geological events during the evolutionary trajectory of South America (Rangel et al. 2018). Finally, we hope that our regionalization scheme stimulates future studies specifically designed to understand a wide range of biogeographical approaches, from internal dynamics at each biogeographic unit (e.g., Godinho and da Silva 2018) to updated revisions of our scheme including newly described species, or implements different metrics (e.g., phylogenetics) to uncover the fascinating mechanisms that led South America to be the most diverse continent in anuran amphibians.

Acknowledgments The authors have been continuously supported by research grants and/or fellowships from the Fundação de Amparo à Pesquisa do Estado de São Paulo (FAPESP 2011/18510-0; 2013/50714-0; 2016/13949-7), Conselho Nacional de Desenvolvimento Científico e Tecnológico (CNPq 2037/2014-9; 431012/2016-4; 308687/2016-17; 114613/2018-4), and University Research and Scientific Production Support Program of the Goias State University (PROBIP/UEG). Prof. Dr. Peter Löwenberg-Neto (UNILA) read critically the first version of this manuscript and provided insightful comments that improved it.

References

Antonelli A, Nylander JAA, Persson C et al (2009) Tracing the impact of the Andean uplift on Neotropical plant evolution. PNAS 106:9749–9754. https://doi.org/10.1073/pnas.0811421106

Antonelli A, Zizka A, Carvalho FA et al (2018) Amazonia is the primary source of Neotropical biodiversity. PNAS 115:6034–6039. https://doi.org/10.1073/pnas.1713819115

Booth BD, Swanton CJ (2002) Assembly theory applied to weed communities. Weed Sci 50:2–13

Borcard D, Legendre P, Drapeau P (1992) Partialling out the spatial component of ecological variation. Ecology 73:1045–1055

Borcard D, Francois G, Legendre P (2011) Numerical ecology with R. Springer, New York

Burnham KP, Anderson DR (2002) Model selection and multimodel inference. Springer, New York

da Silva FR, Almeida-Neto M, Prado VHM et al (2012) Humidity levels drive reproductive modes and phylogenetic diversity of amphibians in the Brazilian Atlantic Forest. J Biogeogr 39:1720–1732

Dapporto L, Ramazzotti M, Fattorini S et al (2013) recluster: an unbiased clustering procedure for beta-diversity turnover. Ecography 36:1070–1075. https://doi.org/10.1111/j.1600-0587.2013.00444.x

Dapporto L, Ciolli G, Dennis RLH et al (2015) A new procedure for extrapolating turnover regionalization at mid-small spatial scales, tested on British butterflies. Methods Ecol Evol 6:1287–1297. https://doi.org/10.1111/2041-210X.12415

Duellman WE, Trueb L (1994) Biology of amphibians. The John Hopkins University Press, Baltimore

Ferro I, Morrone JJ (2014) Biogeographical transition zones: a search for conceptual synthesis. Biol J Linn Soc 113:1–12

Ficetola GF, Mazell F, Thuiller W (2017) Global determinants of zoogeographical boundaries. Nat Ecol Evol 1(4):0089. https://doi.org/10.1038/s41559-017-0089

Godinho MBC, da Silva FR (2018) The influence of riverine barriers, climate, and topography on the biogeographic regionalization of Amazonian anurans. Sci Rep 8:3427. https://doi.org/10.1038/s41598-018-21879-9

Haddad CFB, Toledo LF, Prado CPA et al (2013) Guide to the amphibians of the Atlantic Forest: diversity and biology. Anolis Book, Sao Paulo

Holt BG, Lessard J-P, Borregaard MK et al (2013) An update of Wallace's zoogeographic regions of the world. Science 339:74–78. https://doi.org/10.1126/science.1228282

IBGE (Instituto Brasileiro de Geografia e Estatística) (2012) Manual tecnico da vegetacao brasileira. Instituto Brasileiro de Geografia e Estatística, Rio de Janeiro

Kreft H, Jetz W (2010) A framework for delineating biogeographical regions based on species distribution. J Biogeogr 37:2029–2053. https://doi.org/10.1111/j.1365-2699.2010.02375.x

Lima AP, Magnusson WE, Menin M et al (2006) Guide to the frogs of Reserva Adolpho Ducke, central Amazonia. Attema Design Editorial, Manaus

Lomolino MV, Riddle BR, Whittaker RJ (2017) Biogeography: biological diversity across space and time, 5th edn. Sinauer Associates Inc, Sunderland

Mittermeier RA, Robles-Gil P, Hoffmann M et al (2004) Hotspots revisited: Earths biologically richest and most endangered ecoregions. CEMEX, Mexico City

Morrone JJ (2014) Biogeographical regionalisation of the Neotropical region. Zootaxa 3782:1–110. https://doi.org/10.11646/zootaxa.3782.1.1

Morrone JJ (2015) Biogeographical regionalisation of the Andean region. Zootaxa 3936:207–236. https://doi.org/10.11646/zootaxa.3936.2.3

Morrone JJ (2017) Neotropical biogeography: regionalization and evolution. CRC Press, Taylor & Francis Group, Boca Raton

Olson DM, Dinerstein E, Wikramanayake ED et al (2001) Terrestrial ecoregions of the world: a new map of life on Earth. Bioscience 51:933–938

Pennington RT, Prado DE, Pendry CA (2000) Neotropical seasonally dry forests and Quaternary vegetation changes. J Biogeogr 27:261–273

Proches S (2005) The world's biogeographical regions: cluster analysis based on bat distributions. J Biogeogr 32:607–614. https://doi.org/10.1111/j.1365-2699.2004.01186.x

Rangel TF, Edwards NR, Holden PB et al (2018) Modeling the ecology and evolution of bio-diversity: biogeographical cradles, museums, and graves. Science 361:eaar5452. https://doi.org/10.1126/science.aar5452

Rueda M, Rodríguez MÁ, Hawkins BA (2010) Towards a biogeographic regionalization of the European biota. J Biogeogr 37:2067–2076

Rueda M, Rodríguez MÁ, Hawkins BA (2013) Identifying global zoogeographical regions: lessons from Wallace. J Biogeogr 40:2215–2225

Ruggiero A, Hawkins BA (2008) Why do mountains support so many species of birds? Ecography 31:306–315. https://doi.org/10.1111/j.2008.0906-7590.05333.x

Santos TG, Vasconcelos TS, Rossa-Feres DC et al (2009) Anurans of a seasonally dry tropical forest: Morro do Diabo State Park, São Paulo state, Brazil. J Nat Hist 43:973–993

Southwood TRE (1977) Habitat, the templet for ecological strategies? J Anim Ecol 46:337–365

Valdujo PH, Silvano DL, Colli G et al (2012) Anuran species composition and distribution patterns in the Brazilian Cerrado, a neotropical hotspot. S Am J Herpetol 7:63–78. https://doi.org/10.2994/057.007.0209

Vasconcelos TS, Santos TG, Haddad CFB et al (2010) Climatic variables and altitude as predictors of anuran species richness and number of reproductive modes in Brazil. J Trop Ecol 26:423–432. https://doi.org/10.1017/S0266467410000167

Vasconcelos TS, Rodríguez MÁ, Hawkins BA (2011) Biogeographic distribution patterns of South American amphibians: a regionalization based on cluster analysis. Natureza & Conservação 9:67–72

Vasconcelos TS, Prado VHM, da Silva FR et al (2014) Biogeographic distribution patterns and their correlates in the diverse frog fauna of the Atlantic Forest hotspot. PLoS One 9(8):e104130. https://doi.org/10.1371/journal.pone.0104130

Vilhena DA, Antonelli A (2015) A network approach for identifying and delimiting biogeographical regions. Nat Commun 6:6848. https://doi.org/10.1038/ncomms7848

Villalobos F, Dobrovolski R, Provete DB et al (2013) Is rich and rare the common share? Describing biodiversity patterns to inform conservation practices for South American anurans. PLoS One 8:e56073. https://doi.org/10.1371/journal.-pone.0056073

Wells KD (2007) The ecology and behavior of amphibians. The University of Chicago Press, Chicago

Chapter 7
Spatial Conservation Prioritization for the Anuran Fauna of South America

Abstract South America is the most biologically diverse continent on the planet, including anuran amphibians. However, the continent has been experiencing high levels of habitat degradation, and amphibians are considered the most endangered class of the vertebrate group globally. Therefore, the establishment of effective actions for the anuran protection in the continent is urgent. Here, we generate a spatial conservation prioritization of anurans in South America addressing different human-related and biological diversity metrics using the software MARXAN. We found that the anuran fauna of South America can be totally represented by the selection of ~19.53% of the total area of the continent. Contiguous selected areas are mainly located in the Tropical Andes and the Atlantic Forest coast. To a lesser extent, the selected areas can be also found in specific areas of Venezuela, the Brazilian Amazonian forest, and the temperate Chilean forests. In general, the contiguous areas represent forested areas within rough topographies of tropical countries and should be priority areas for anuran conservation in South America. Other areas within the tropical region, as well as in the southern temperate regions/countries, are less continuous and should involve more complex evaluations by decision-makers to foster reserve creation.

Keywords Anura · Biological conservation · Conservation biogeography · MARXAN · Neotropics · Spatial prioritization

7.1 Introduction

Our planet is currently experiencing biodiversity loss and degradation of ecosystems at rates never recorded in our evolutionary history (Barnosky et al. 2011). The drivers of this biodiversity crisis (e.g., habitat change, invasive alien species, climate change, overexploitation, and pollution) are expected to remain either steady or increase in intensity in future scenarios (Millennium Ecosystem Assessment 2005). The situation in South America is especially worrisome, since

© Springer Nature Switzerland AG 2019
T. S. Vasconcelos et al., *Biogeographic Patterns of South American Anurans*,
https://doi.org/10.1007/978-3-030-26296-9_7

the continent is one of the most biodiverse on Earth, yet has also experienced some of the most intense habitat loss rates (Whitmore 1997; Primack et al. 2001; UNEP-WCMC 2016).

Amphibians are the most threatened terrestrial vertebrate group globally (Catenazzi 2015; IUCN 2019). The main threats to amphibians are generally related to habitat loss and fragmentation, fungal disease, and climate change, among others (e.g., Catenazzi 2015; Scheele et al. 2019). Effective actions to protect these animals are important because they play key important roles in both aquatic and terrestrial ecosystems that will ultimately be offered to humans as regulatory and/or provisioning ecosystem services. For instance, anurans provide biological control of insect populations, and their chemically diverse skin secretions can be used to derive pharmacological products (Duellman and Trueb 1994; Stebins and Cohen 1995). Although South America supports the highest anuran diversity in the world (Wake and Koo 2018; Chap. 2 of this book), current drivers of biodiversity crisis also threaten anurans in this continent (e.g., Becker et al. 2007; Dixo and Metzger 2010; Becker et al. 2015; Bovo et al. 2016; Vasconcelos and Doro 2016; Vasconcelos et al. 2018). Therefore, effective conservation actions are urgent to protect anurans in the continent.

An important step for delineating conservation strategies is understanding species distribution dynamics at biogeographical scales (Whittaker et al. 2005; Ladle and Whittaker 2011). Conservation biogeography is an interdisciplinary field with tools and approaches from populations to ecosystems and landscape to broader scales (Ladle and Whittaker 2011). Among these different approaches, the process of systematic conservation planning aims to identify conservation goals for a given region, review existing conservation areas, and implement effective conservation actions. Within this process, there is an important step aimed to select biologically important areas at a minimum set coverage, the process of spatial conservation prioritization (Margules and Pressey 2000). This process begins with a matrix of species distribution within an area of interest that is submitted to optimization procedures in order to find a minimum cover of the biodiversity target (e.g., composition and representation of all species, structure of habitat types, or the functioning of ecological processes) (Ardron et al. 2010; Kukkala and Moilanen 2013 and references therein). Therefore, the selection of biologically representative areas is an important step to support decision-making for designing reserve network.

In this chapter, we generate a spatial conservation prioritization of anurans in South America. Though previous spatial prioritization proposals have been performed in South America, they were either directed toward broader geographic scales within specific aims or considered amphibians jointly with other taxa (e.g., mammals, birds, and amphibians: Moilanen et al. 2013; Durán et al. 2014). Here, we generate a spatial prioritization with the updated dataset of South American anurans detailed in Chap. 2. In order to improve our prioritization scheme, we also addressed different human-related and biological diversity metrics important in assisting effective conservation proposals, due to the multifaceted concept of biological diversity and the different human impacts that may difficult conservation efforts (e.g., Carvalho et al. 2011; Lagabrielle et al. 2018; Vasconcelos and Prado 2019). Finally, our conservation proposal is intended to represent areas with high anuran species and/or support unique species composition, as well as areas having the highest phylogenetic diversity and lower levels of anthropogenic impacts.

7.2   Material and Methods

The presence/absence matrix of anurans within the grid system of South America, detailed in Chap. 1, was submitted to the spatial conservation prioritization using MARXAN (Ball and Possingham 2000; Ball et al. 2009). This software uses a simulated annealing algorithm that is ultimately aimed to find good solutions to the minimum set problem (Ardron et al. 2010). That is, the algorithm selects sampling units (e.g., grid cells) to represent the biodiversity targets at a minimum possible cost, so it starts with a random set of grid cells that are replaced after a number of iterations. Iteration by iteration, the algorithm either keeps good solutions or replaces grid cells if the selection of new sampling units improves the optimization process to finally find an efficient spatial solution (Ardron et al. 2010). Here, our biodiversity target is to find minimum representations of anurans across the continent, so a combination of grids containing either high species richness or complementary species from one another is expected in the final spatial conservation prioritization. The prioritization scheme also considered the conservation cost of each grid cell based on human impacts (e.g., Vasconcelos and Prado 2019). To do so, we considered the human footprint index (WCS and CIESIN 2005) to represent a range of anthropogenic impacts (e.g., density of human population, concentration of urban areas, roads, agricultural land uses, and navigable rivers) that may be difficult conservation efforts (e.g., Lagabrielle et al. 2018; see also Chap. 1 for detailed source information and data handling of the human footprint index). Then, higher values of human footprint represent highly altered areas difficult to be recovered, being thus less prioritized in the final prioritization solution (e.g., Vasconcelos and Prado 2019). The optimization process also considered those grid cells with high phylogenetic diversity. Therefore, we use grid cells with the upper 25% percentile of phylogenetic diversity identified in Chap. 3 to be prioritized in our conservation proposal.

In order to obtain a compacted/connected spatial solution rather than a patchy distribution of prioritized cells, we set the boundary length modifier (BLM) to one (Ardron et al. 2010). The optimization procedure was repeated 100 times, and the most frequently selected grid cells were identified using the metric "selection frequency." Since the different MARXAN runs generate variable network solutions, the "selection frequency" function enables us to recognize those most efficient grid cells identified among the 100 runs. Then, our final prioritization scheme is based on grid cells selected in more than 50% of the runs due to their efficiency in meeting biodiversity goals (Ardron et al. 2010; Vasconcelos and Prado 2019).

7.3   Results and Discussion

The anuran fauna of South America can be totally represented by the selection of 322 grid cells, of which accounts for 19.53% of the total area of the continent (Fig. 7.1). Overall, contiguous selected areas are mainly located in the Tropical Andes and the Atlantic Forest coast (Fig. 7.1). To a lesser extent, the selected areas can be also found in specific areas of Venezuela, the Brazilian Amazonian forest, and the temperate Chilean forests (Fig. 7.1).

The present study broadly corroborates previous spatial conservation prioritizations that used different vertebrate taxa as biological models (amphibians, birds, and mammals; Moilanen et al. 2013; Dobrovolski et al. 2014; Durán et al. 2014). These previous studies and our results point out that considerable areas within Colombia,

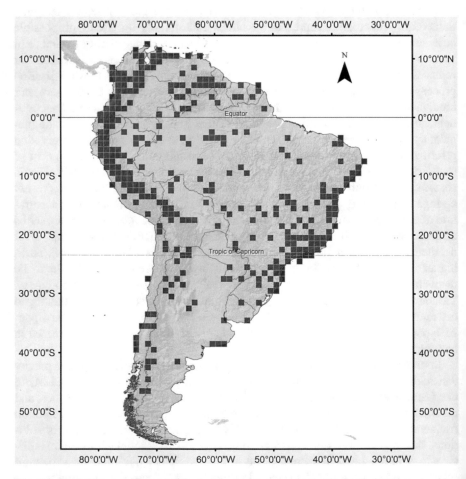

Fig. 7.1 Spatial conservation prioritization for anuran conservation in South America, considering species composition, phylogenetic diversity, and the human footprint index (see Methods). Red cells indicate sites selected by the MARXAN algorithm as priority for conservation

Ecuador, the Atlantic Forest coast in Brazil, and Peru are of biological relevance for biodiversity conservation. The common feature of these selected areas is the rough topography and/or forested habitat. Both such mountainous and forested areas encompass three biologically rich, yet deeply threatened, places recognized by Mittermeier et al. (2004) as biodiversity hotspots: the Brazilian Atlantic Forest, the Tropical Andes, and the Tumbes-Chocó-Magdalena. Moreover, these areas also support high levels of taxonomic (Chap. 3), phylogenetic (Chap. 4), and functional (Chap. 5) diversity of anurans.

The recognition of biologically important areas for anuran conservation in Brazil, Colombia, Ecuador, and Peru does not decrease the biological value of other areas in South America. As recognized in Chap. 6, different species pools are found across the continent. For instance, the Amazon biogeographic region is relatively under-represented by our prioritization protocol. Nonetheless, we found previously (Chaps. 3, 4, and 5) that this area supports one of the highest levels of taxonomic, phylogenetic, and functional diversity in the continent. Additionally, even species-poor countries and/or biomes in South America harbor endemic species that represent different evolutionary lineages, such as the hotspot Chilean Winter Rainfall and Valdivian forests, as well as the Cerrado hotspot (Mittermeier et al. 2004; Valdujo et al. 2012). These areas were represented in our prioritization proposal, though they are less abundant and patchier distributed than those selected in the tropical mountainous and forested areas. Thus, our prioritization scheme not only identified major areas of biological relevance for anuran conservation but also specific grid cells in which systematic conservation planners may want to take a closer look to devise specific conservation actions, which may be the case of those countries and biomes that were underrepresented by our prioritization scheme.

In summary, we found that forested areas within rough topographies of tropical countries should be priority areas for anuran conservation in South America. Other areas within the tropical region, as well as in the southern temperate regions/countries, also have biologically important areas. However, these areas are less continuous and more disconnected, which in turn involve more complex evaluations by decision-makers to foster reserve creation. Finally, we emphasize that our prioritization scheme is not definitive, since new anuran species are being described at high rates decade by decade (see Chap. 2). Moreover, the intensity of threats to biodiversity may change over time, so as soon as these threats are mapped for the whole continent (e.g., Garcia et al. 2014), they can be implemented in optimization processes to generate updated and more accurate spatial conservation prioritizations. Irrespective, in light of such intense threats of the current biodiversity crisis, our prioritization proposal may be an important guide for herpetologists and conservation biogeographers for supporting effective conservation actions for the protection of South American anurans.

Acknowledgments The authors have been continuously supported by research grants and/or fellowships from the Fundação de Amparo à Pesquisa do Estado de São Paulo (FAPESP 2011/18510-0; 2013/50714-0; 2016/13949-7), Conselho Nacional de Desenvolvimento Científico e Tecnológico (CNPq 2037/2014-9; 431012/2016-4; 308687/2016-17; 114613/2018-4), and University Research and Scientific Production Support Program of the Goias State University (PROBIP/UEG).

References

Ardron JA, Possingham HP, Klein CJ (2010) Marxan good practices handbook. Version 2. Pacific Marine Analysis and Research Association, Victoria

Ball IR, Possingham HP (2000) Marxan (v 1.8.6): marine reserve design using spatially explicit annealing. A manual prepared for the Great Barrier Reef Marine Park Authority

Ball IR, Possingham HP, Watts M (2009) Marxan and relatives: software for spatial conservation prioritisation. In: Moilanen A, Wilson KA, Possingham HP (eds) Spatial conservation prioritisation: quantitative methods and computational tools. Oxford University Press, Oxford, UK, pp 185–195

Barnosky AD, Matzke N, Tomiya S et al (2011) Has the Earth's sixth mass extinction already arrived? Nature 471:51–57. https://doi.org/10.1038/nature09678

Becker CG, Fonseca CR, Haddad CFB et al (2007) Habitat split and the global decline of amphibians. Science 318:1775–1777

Becker CG, Rodriguez D, Lambertini C et al (2015) Historical dynamics of Batrachochytrium dendrobatidis in Amazonia. Ecography 39:954. https://doi.org/10.1111/ecog.02055

Bovo RP, Andrade DV, Toledo LF et al (2016) Physiological responses of Brazilian amphibians to an enzootic infection of the chytrid fungus *Batrachochytrium dendrobatidis*. Dis Aquat Org 117:245–252

Carvalho SB, Britto JC, Crespo EJ et al (2011) Incorporating evolutionary processes into conservation planning using species distribution data: a case study with the western Mediterranean herpetofauna. Divers Distrib 17:408–421. https://doi.org/10.1111/j.1472-4642.2011.00752.x

Catenazzi A (2015) State of the World's amphibians. Annu Rev Environ Resour 40:91–119

Dixo M, Metzger JP (2010) The matrix-tolerance hypothesis? An empirical test with frogs in the Atlantic Forest. Biodivers Conserv 19:3059–3071

Dobrovolski R, Loyola R, da Fonseca GAB et al (2014) Globalizing conservation efforts to save species and enhance food production. Bioscience 64:539–545. https://doi.org/10.1093/biosci/biu064

Duellman WE, Trueb L (1994) Biology of amphibians. John Hopkins University Press, Baltimore

Durán AP, Duffy JP, Gaston KJ (2014) Exclusion of agricultural lands in spatial conservation prioritization strategies: consequences for biodiversity and ecosystem service representation. Proc Biol Sci 281:20141529

Garcia RA, Cabeza M, Rahbek C et al (2014) Multiple dimensions of climate change and their implications for biodiversity. Science 344:1247579. https://doi.org/10.1126/science.1247579

IUCN (2019) *The IUCN Red List of Threatened Species. Version 2019-1*. http://www.iucnredlist.org. Accessed 21 Mar 2019

Kukkala AS, Moilanen A (2013) Core concepts of spatial prioritization in systematic conservation planning. Biol Rev Camb Philos 88:443–464. https://doi.org/10.1111/brv.12008

Ladle RJ, Whittaker RJ (2011) Conservation biogeography. Wiley-Blackwell, West Sussex

Lagabrielle E, Lombard AT, Harris JM et al (2018) Multi-scale multi-level marine spatial planning: a novel methodological approach applied in South Africa. PLoS One 13:e0192582. https://doi.org/10.1371/journal.pone.0192582

Margules CR, Pressey RL (2000) Systematic conservation planning. Nature 405:243–253

Millenium Ecosystem Assessment (2005) Ecosystems and human well-being: biodiversity synthesis. World Resources Institute, Washington, DC

Mittermeier RA, Robles-Gil P, Hoffmann M et al (2004) Hotspots revisited: Earths biologically richest and most endangered ecoregions. CEMEX, Mexico City

Moilanen A, Anderson BJ, Arponen A et al (2013) Edge artefacts and lost performance in national versus continental conservation priority areas. Divers Distrib 19:171–183

Primack R, Rozzi R, Massardo F, Feinsinger P (2001) Destrucción y degradación del habitat. In: Primack R, Rozzi R, Massardo F, Feinsinger P (eds) Fundamentos de conservación biológica: perspectivas latinoamericanas. Fondo de Cultura Económica, México, pp 183–223

Scheele BC, Pasmans F, Skerratt LF et al (2019) Amphibian fungal panzootic causes catastrophic and ongoing loss of biodiversity. Science 363:1459–1463. https://doi.org/10.1126/science.aav0379

Stebins RC, Cohen NW (1995) A natural history of amphibians. Princeton University Press, New Jersey

UNEP-WCMC (2016) The State of Biodiversity in Latin America and the Caribbean: a mid-term review of progress towards the Aichi Biodiversity Targets. UNEP-WCMC, Cambridge, UK. Available at: https://www.cbd.int/gbo/gbo4/outlook-grulac-en.pdf

Valdujo PH, Silvano DL, Colli G et al (2012) Anuran species composition and distribution patterns in the Brazilian Cerrado, a neotropical hotspot. S Am J Herpetol 7:63–78. http://www.bioone.org/doi/full/10.2994/057.007.0209

Vasconcelos TS, Doro JLP (2016) Assessing how habitat loss restricts the geographic range of Neotropical anurans. Ecol Res 31:913–921

Vasconcelos TS, Prado VHM (2019) Climate change and opposing spatial conservation priorities for anuran protection in the Brazilian hotspots. J Nat Conserv 49:118–124. https://doi.org/10.1016/j.jnc.2019.04.003

Vasconcelos TS, Nascimento BTM, Prado VHM (2018) Expected impacts of climate change threaten the anuran diversity in the Brazilian hotspots. Ecol Evol 2018(8):7894–7906. https://doi.org/10.1002/ece3.4357

Wake DB, Koo MS (2018) Primer: amphibians. Curr Biol 28:R1221–R1242

Whitmore TC (1997) Tropical forest disturbance, disappearance, and species loss. In: Laurance WF, Bierregaard RO Jr (eds) Tropical forest remnants: ecology, management, and conservation of fragmented communities. The University of Chicago Press, Chicago, pp 3–12

Whittaker RJ, Araújo MB, Jepson P et al (2005) Conservation biogeography: assessment and prospect. Divers Distrib 11:3–23

Wildlife Conservation Society - WCS, Center for International Earth Science Information Network - CIESIN - Columbia University (2005) Last of the Wild Project, Version 2, 2005 (LWP-2): Global Human Footprint Dataset (Geographic)

Index

© Springer Nature Switzerland AG 2019
T. S. Vasconcelos et al., *Biogeographic Patterns of South American Anurans*,
https://doi.org/10.1007/978-3-030-26296-9

Printed in the United States
By Bookmasters